D1156982

POWER PLANT ENTRAINMENT
A Biological Assessment

ACADEMIC PRESS RAPID MANUSCRIPT REPRODUCTION

POWER PLANT ENTRAINMENT

A Biological Assessment

EDITED BY

J. R. Schubel

Marine Sciences Research Center
State University of New York
Stony Brook, New York

Barton C. Marcy, Jr.

Ecological Sciences Division
NUS Corporation
Pittsburgh, Pennsylvania

ACADEMIC PRESS New York San Francisco London 1978
A Subsidiary of Harcourt Brace Jovanovich, Publishers

ACADEMIC PRESS, INC.
111 Fifth Avenue, New York, New York 10003

United Kingdom Edition published by
ACADEMIC PRESS, INC. (LONDON) LTD.
24/28 Oval Road, London NW1 7DX

Library of Congress Cataloging in Publication Data

Main entry under title:

Power plant entrainment.

A summary of the workshop held by the Ad Hoc Committe on Entrainment on Jan. 17-21 and Mar. 20-22, 1977 at the Marine Sciences Research Center, State University of New York at Stony Brook.

Includes index.

1. Fishes, Effect of water pollution on—Congresses. 2. Electric power-plants—Environmental aspects—Congresses. 3. Electric power-plants—Design and construction—Congresses. 4. Water quality—Standards—Congresses. I. Schubel, J. R. II. Marcy, Barton C.

SH177.E4P67 628.1'683 78-1509

ISBN 0-12-631050-5

PRINTED IN THE UNITED STATES OF AMERICA

CONTENTS

THE AD HOC COMMITTEE ON ENTRAINMENT

ALLAN D. BECK
U.S. Environmental Protection Agency
Environmental Research Laboratory
Narragansett, RI 02882

EDWARD J. CARPENTER
Marine Sciences Research Center
State University of New York
Stony Brook, NY 11794

CHARLES C. COUTANT
Environmental Sciences Division
Oak Ridge National Laboratory
Oak Ridge, TN 37830

BLAIR KINSMAN
Marine Sciences Research Center
State University of New York
Stony Brook, NY 11794

BARTON C. MARCY, JR.
Ecological Sciences Division
NUS Corporation
Pittsburgh, PA 15220

RAYMOND P. MORGAN III
Chesapeake Biological Laboratory
University of Maryland
Solomons, MD 20688

ALAN S. ROBBINS
Marine Sciences Research Center
State University of New York
Stony Brook, NY 11794

J. R. SCHUBEL
Marine Sciences Research Center
State University of New York
Stony Brook, NY 11794

ROBERT E. ULANOWICZ
Chesapeake Biological Laboratory
University of Maryland
Solomons, MD 20688

P. M. J. WOODHEAD
Marine Sciences Research Center
State University of New York
Stony Brook, NY 11794

PREFACE

In 1976 we formed an ad hoc Committee on Entrainment because of our growing discontent with the failure of decision-makers to set, for power plants with once-through cooling systems, appropriate design and operating criteria, criteria to minimize entrainment losses. On 17–21 January 1977, and again on 20–22 March 1977, this committee met at the Marine Sciences Research Center of the State University of New York at Stony Brook to make a critical assessment of the effects of entrainment by power plants with once-through cooling systems. This report is a summary of the results of that workshop.

The primary goals of the workshop were

(1) to assess the effects of the several stresses associated with entrainment—thermal, physical, and chemical,

(2) to use the results of this assessment to develop guidelines for the conceptual design and operation of power plants with once-through cooling systems to minimize the mortalities of entrained organisms, and to minimize the effects on the populations of those organisms,

(3) to outline research priorities required: (a) for a significant improvement in our ability to make an unequivocal assessment of the mortalities associated with entrainment; (b) to separate the effects of each of the several stresses; and (c) to improve the design and operating criteria of power plants with once-through cooling systems to ensure acceptable and predictable mortalities due to entrainment.

We have not considered far-field thermal effects—effects on the population and community—that might result from cropping by entrainment or from small persistent increases in temperature over relatively large areas of the receiving water body. A recent conference dealt with these topics.* Nor have we considered problems of impingement and entrapment of fish on intake screens. We have also elected to neglect consideration of any effects that might accrue from the discharge of metals and radioactivity to the aquatic environment by once-through cooling systems. Temperature is considered only as it is a fac-

*Van Winkle, W., ed., *Proc. Conf. Assessing Effects of Power-Plant Induced-Mortality on Fish Populations, Gatlinburg,* Tennessee, May 3–6. Pergamon Press, New York, 1977.

tor in the entrainment process, not in its potential for effects in the receiving waters.

This report was written primarily for decision makers and their staffs in state and federal regulatory agencies responsible for setting design and operating criteria and for assessing the impacts of power plants with once-through cooling systems. It will be useful to managers of funding agencies that support power plant related research, to scientists who conduct the research, and to a variety of other groups. We hope it will also be studied by environmentalists, particularly those who, out of a concern about possible undesirable effects of "thermal pollution," have pressed for stringent thermal regulations for power plants with once-through cooling systems, often unaware of the concomitant increase in risks to organisms from physical factors during entrainment.

While most of the chapters have been written by one or several committee members, the complete document has been read and endorsed by the entire committee. Chapter 7 was written by the Committee at large. Because each chapter was intended to stand alone, there is some redundancy. This was inevitable and, in our view, desirable.

The workshop and preparation of this report were directly supported by the New York State Energy Research and Development Authority, the New York Sea Grant Institute, the Marine Sciences Research Center of the State University of New York, the NUS Corporation, and the Rockefeller Foundation. Salary support was generously provided by each of the participant's employers.

Karen Chytalo, Jack Lekan, Norman Itzkowitz, Christopher Smith, and Alexis Steen assisted in a variety of ways during the workshop. Editorial help was provided by Luise Davis and Richard Nugent of the Ecological Sciences Division of NUS Corporation, Pittsburgh, Pennsylvania. We are indebted to Phillip Goodyear, Gary Milburn, and Joseph O'Connor for reading the entire manuscript and for their helpful suggestions. We thank Bart Chezar of the New York State Energy Research and Development Authority and Harry H. Carter of the Marine Sciences Research Center for encouragement and advice. We thank Jeri Schoof, Luise Davis, Vera Percy, and Alice Lawson for taking the manuscript from rough draft to final copy, and for their manifold contributions that were so important to the completion of this project.

CHAPTER 1. INTRODUCTION

THE COMMITTEE ON ENTRAINMENT

TABLE OF CONTENTS

I. SOME CHARACTERISTICS OF STEAM ELECTRIC PLANTS

A steam electric plant typically consists of a heat source, boiler, turbine, generator, and a condenser system, Fig. 1. Steam from the boiler drives the turbine which spins the generator to produce electricity. After passing through the turbine, the steam must be condensed and returned to the boiler. The most economical

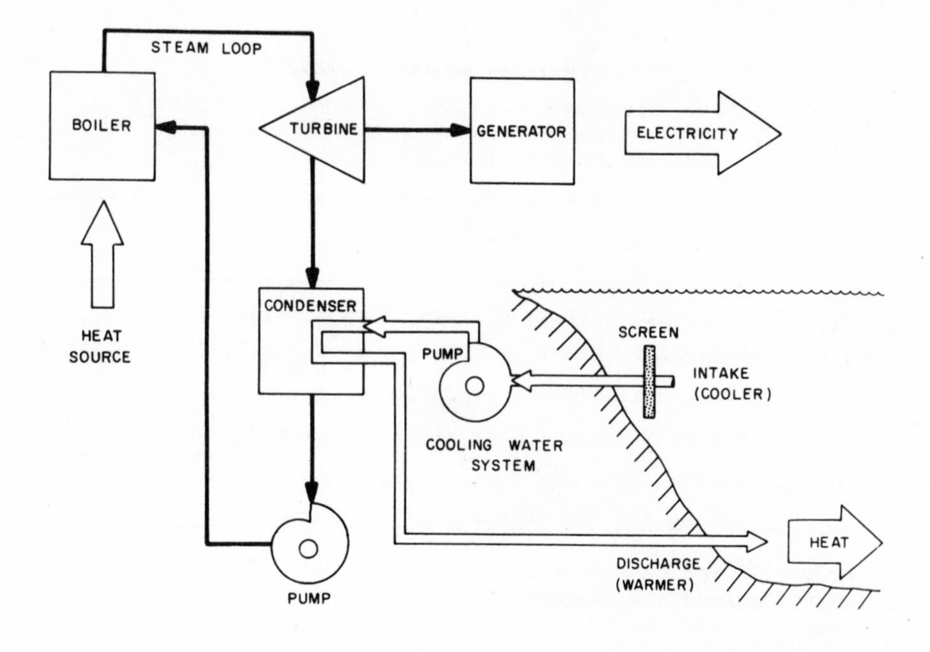

Fig. 1. Schematic of a steam electric plant with a once-through cooling system.

way of achieving this is a once-through, or open-cycle, cooling system which passes water from the environment through the condenser system and discharges it back into the environment at an elevated temperature.

In 1973, the latest year for which there are published data, 60% of the 769 operating fossil fuel steam electric plants in the United States used once-through cooling; and 74% of the 39 operating nuclear power plants had once-through cooling systems, Tables 1 and 2.

TABLE 1[a] Frequency of Occurrence (%) of Different Types of Cooling Systems Used by U.S. Operating Fossil Fuel Power Plants, 1969-1973

Cooling System	*1969*	*1970*	*1971*	*1972*	*1973*
Once-through (fresh water)	*49.8*	*49.9*	*48.1*	*47.2*	*44.0*
Once-through (saline water)	*18.9*	*18.4*	*18.1*	*17.3*	*16.4*
Cooling Ponds	*5.4*	*5.6*	*6.0*	*6.3*	*6.0*
Cooling Towers	*17.2*	*17.3*	*18.1*	*18.6*	*21.5*
Combined Systems	*8.7*	*8.8*	*9.7*	*10.6*	*12.1*
TOTAL (%)	*100.0*	*100.0*	*100.0*	*100.0*	*100.0*

[a]*Federal Power Commission, 1976. Steam-Electric Plant Air Quality Control Data, Summary Report for the Year Ended December 31, 1973 based on FPC form No. 67. Federal Power Commission, Washington, D.C., FPC-S-253, January 1976.*

Although once-through cooling is the most economical way of condensing the exhaust steam from the turbines of steam electric plants, the volumes of water used for this purpose and the quantities of "waste" heat added to the aquatic environment are extremely large, thus prompting demands for alternatives. The total of approximately 500 steam electric plants with once-through cooling systems operating in the United States in 1973 removed

TABLE 2[a] Frequency of Occurrence (%) of Different Types of Cooling Systems Used by U.S. Operating Nuclear Power Plants, 1969-1973

Cooling System	1969	1970	1971	1972	1973
Once-through (fresh water)	58.3	56.3	52.4	48.1	51.2
Once-through (saline water)	25.0	25.0	23.8	29.6	23.1
Cooling Ponds	16.7	12.4	14.2	11.2	7.7
Cooling Towers	--	--	4.8	3.7	5.2
Combined Systems	--	6.3	4.8	7.4	12.8
TOTAL	100.0	100.0	100.0	100.0	100.0

[a]*Alexis Steen, personal communication, January 1977. Data from "Assessment of Water Use by Nuclear Power Plants," memo Specialists Branch, U.S. Nuclear Regulatory Commission, 13 May 1975.*

water from a variety of water bodies at a combined rate of more than 8900 m^3/s, Table 3. This flow is equivalent to about 13% of

TABLE 3[a] Rates of Water Withdrawal (m^3/s) by U.S. Fossil Fuel and Nuclear Steam Electric Plants Between 1969 and 1973

Type of Plant	1969	1970	1971	1972	1973
Fossil Fuel Plants					
Fresh water	4679	4871	4882	5169	5532
Saline water	1937	2080	2055	2128	2187
Nuclear Plants					
Fresh water	145	219	363	418	846
Saline water	106	133	133	276	370
TOTAL (m^3/s)	6867	7303	7433	7991	8935

[a]*Federal Power Commission, 1976. Steam-Electric Plant Air and Water Quality Control Data, Summary Report for the Year Ended December 31, 1973 based on FPC form No. 67. Federal Power Commission, Washington, D.C., FPC-S-253, January 1976.*

the annual average discharge of all rivers and streams in the coterminous United States. This is not a consumptive use, however. Most of the water that is withdrawn is returned to the source water bodies, but after being heated by from 5 to about 40°C, after chlorination (usually periodically), and after passage through the intake screens and the cooling circuit.

The range in the temperature rise across the condensers (ΔT) of operating power plants with once-through cooling systems is from 5 to 40°C. A histogram of the ΔT's of operating and proposed nuclear plants as of 1976 is presented in Fig. 2. Similar data

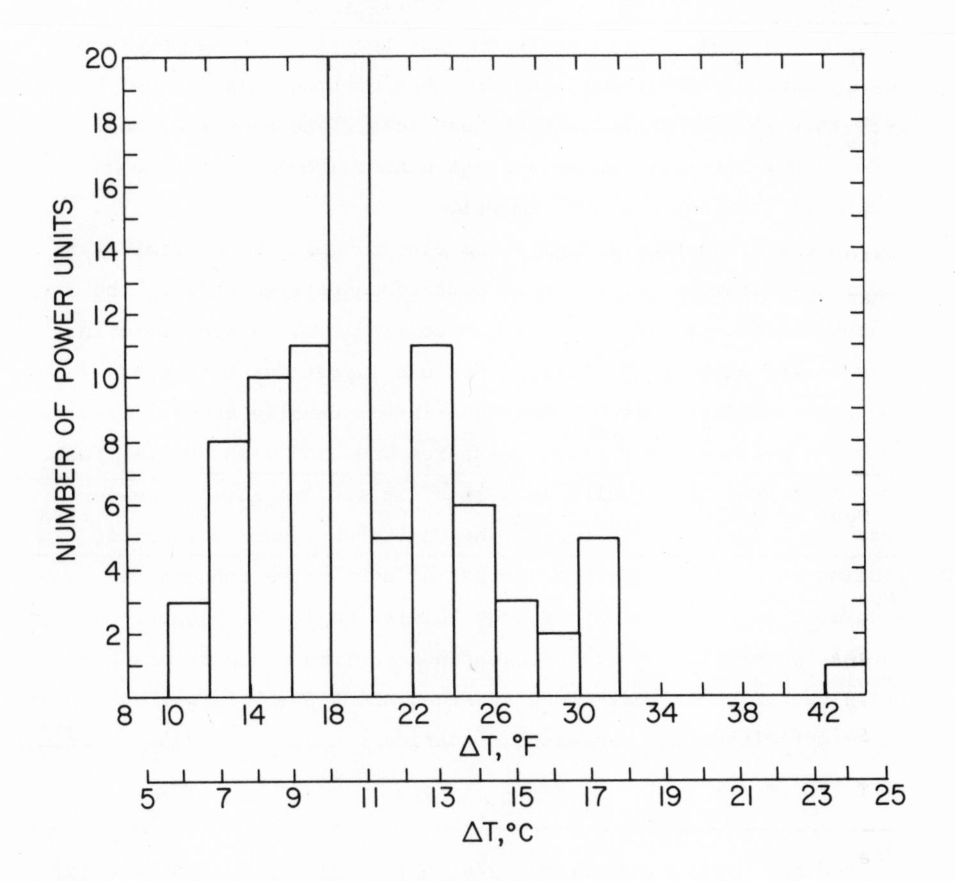

Fig. 2. Histogram of ΔT's at proposed and operating nuclear power plants (1976).

from all fossil fuel plants are not readily available.

The quantity of waste heat rejected to the environment by a steam electric plant depends upon the type of plant and its efficiency. The most efficient plants achieve efficiencies of about 40%, and the average for all steam electric plants operating in 1973 was about 33%. In the operation of a plant, some heat is lost within the plant and through the stack, but on the average, more than 50% of the heat input is transferred to the cooling water in the steam condensing process. In 1973, the total amount of heat rejected to the aquatic environment by once-through cooling systems of steam electric plants exceeded 1.6×10^{18} gm cal. This is enough heat to raise the temperature of a volume of water equivalent to that of Long Island Sound by more than 25°C if there were no heat loss. The amount of heat used by New York City in an average winter (November through February) is about 2×10^{17} gm cal.

Nuclear and fossil fuel steam electric plants are similar in their operation. The principal difference between them is the energy source. Their once-through cooling systems are alike in essentials, although nuclear plants use larger quantities of cooling water per unit of power generated, usually about 1.5 times as much, because lower steam pressures are used with nuclear fuel. The lower pressure results in less efficient operation (greater heat rejection) and, therefore, necessitates larger volumes of cooling water to maintain a desired ΔT across the condensers. According to a U.S. Atomic Energy Commission (1969) report, nuclear power plants with once-through cooling systems, planned or in operation at that time, averaged about 3 m^3 of cooling water per minute per megawatt (electrical).

II. ENTRAINMENT

Not only water is passed through the once-through cooling systems of steam electric plants, but a variety of organisms as well. Plankton and many weak swimming nekton with maximum cross-sectional dimensions smaller than the mesh size[1]--generally 9 to 13 mm--of the screens covering the cooling water intakes of steam electric plants may be carried along with the cooling water through the plant and be subjected to the accompanying thermal stress and to a variety of physical and chemical stresses. Other organisms in the receiving waters may be entrained into the effluent along with the diluting water without passing through the plant.

We define *entrainment* as the capture and inclusion of organisms in the cooling water of power plants. There are two modes of entrainment: *pump (or intake) entrainment* in which organisms are pumped through the plant and discharged back into the environment, and *plume entrainment* in which organisms are incorporated into the discharge plume along with diluting water without having passed through the plant.

Organisms too large to pass through the intake screens may be drawn into the intakes with the cooling water and become *entrapped* on the screens and injured or killed by *impingement*. In some cases, this type of damage may contribute significantly to the total mortality of organisms affected by a plant's cooling water system.

All planktonic (drifting) organisms and some nektonic (swimming) organisms are susceptible to *plume* entrainment. Those

[1]*Since many fish and other aquatic organisms are elongated, relatively large organisms may pass through a plant's cooling system; entrained larval and juvenile finfish of up to 50 or 60 mm in length are not uncommon.*

planktonic and nektonic organisms with cross-sectional dimensions less than the mesh size of a plant's intake screens are susceptible to *pump* entrainment. Entrained organisms thus range from microscopic bacteria and viruses to relatively large weak swimming macroscopic organisms. Occasionally, when the discharge velocities are high, even strong swimming adult fish may be entrained into the plume of discharge water.

III. THE PERCEIVED EFFECTS OF ENTRAINMENT AND THE ADOPTION OF THERMAL REGULATIONS

The potential undesirable effects of the thermal stresses associated with power plants with once-through cooling systems led many biologists and environmentalists to campaign vigorously against "thermal pollution" of the aquatic environment. The intensity and persistence of these concerns resulted in the adoption, by the Federal government and by various states, of stringent thermal regulations to mitigate the anticipated adverse thermal effects associated with once-through cooling systems of steam electric plants. These regulations typically have taken the form of limiting any or all of the following: the maximum temperature rise across the condensers (the ΔT), the maximum temperature of the effluent, the surface area of a mixing zone in the receiving waters within which some specified temperature could be exceeded, and the maximum temperature of the mixed water body outside the mixing zone. These regulations have been met by using relatively low ΔT's across the condensers of many steam electric stations with once-through cooling systems.

Since a plant must reject heat to the environment at a fixed rate, the smaller the temperature rise across the condensers, the larger the volume of water that must be pumped to achieve the required cooling. While this practice may minimize mortalities resulting from *thermal* stresses experienced by organisms carried through a plant, or entrained into the heated discharge, it ignores the effects of chemical stresses associated with biocides used to prevent fouling of the cooling system and the physical stresses associated with pressure changes, shear forces, impact, and abrasion during passage through the cooling circuit. Initial assessments of the mortalities of organisms carried through the once-through cooling systems of steam electric plants focused on thermal impacts. Recently, however, consideration has been given to assessing the effects of chemical and physical stresses as well.

Once-through systems as a mode of cooling for steam electric stations have now been indicted as environmentally degrading by many, and alternate methods of cooling have been sought. Cooling towers using closed-cycle cooling are viewed by many as a panacea to the problems of heat rejection to the environment. Although cooling towers are not without their environmental and economic drawbacks, recent legislation seems certain to increase the use of closed-cycle (cooling towers) cooling over open-cycle (once-through) cooling. In fact, as the law stands some existing plants with open-cycle cooling may be required to backfit for cooling towers. Other types of "closed-cycle" cooling systems include cooling ponds (lakes or reservoirs), and spray ponds or canals.

IV. THE LAWS

The National Environmental Policy Act of 1969 (NEPA) PL 91-190 was a major attempt to focus and coordinate all federal actions that are "significantly affecting the quality of the human environment" to the need for detailed environmental analyses (National Environmental Policy Act, 1969). The Council on

Environmental Quality (CEQ) under Title II of NEPA, is the administrator and clearinghouse for the environmental impact statements from the involved Federal agencies. NEPA guidelines were published in the Code of Federal Regulations, Title 40, Chapter V, Part 1500, and appeared in the Federal Register, August 1, 1973, pp. 20549-20562. The NEPA guidelines provide a mechanism whereby Federal agencies must assess fully the potential environmental impact of a proposed action as early as possible. In all cases, the assessment must be made prior to agency decision concerning legislative actions which may significantly affect the environment. A major landmark for power plant impact assessment was the Calvert Cliffs decision (*Calvert Cliffs Coordinating Comm., et al., vs U.S. Atomic Energy Commission,* 1971) by the U.S. Court of Appeals for the District of Columbia which mandated thorough review of thermal effects of nuclear power plants.

The Federal Water Pollution Control Act Amendments of 1972 (FWPCA) PL 92-500 were a major package of comprehensive legislation following soon after NEPA. The FWPCA proclaimed two national goals:

(1) To eliminate the discharge of pollutants into navigable waters by 1985,

(2) To achieve wherever attainable, an interim goal of water quality which provides for the protection and propagation of fish, shellfish, and wildlife and provides for recreation in and on the water by July 1, 1983.

With the passage of the Act, a national program was implemented which requires a permit for every point source discharge of pollutants--including waste heat--into the navigable waters of the United States. This program is known as the National Pollutant Discharge Elimination System (NPDES), Section 402. A state may acquire and administer the NPDES program if it applies to the EPA and meets the EPA's legal requirements under "State Program Elements Necessary for Participation in the National

Pollutant Discharge Elimination System: 40 CFR Part 124." If a state does not meet the requirements of 40 CFR Part 124, then EPA retains the permit authorization. Section 301(b)(1)(A) of the Act requires all point sources, other than publicly owned treatment works, to achieve the "best practicable control technology currently available" by 1 July 1977 and the "most stringent best available technology economically achievable by no later than 1 July 1983."

Section 316(a) of the Act provides an opportunity for the owner, or operator, of a steam electric generating station to demonstrate that the *thermal* effluent limitations imposed by the Administrator of the EPA (or, if appropriate, the State) pursuant to Section 301 of the Act are more stringent than necessary to

> ". . . assure the protection and propagation of a balanced, indigenous population of shellfish, fish, and wildlife in and on the body of water into which the discharge is to be made, . . ."

If a successful demonstration is made, the Administrator of the EPA (or, if appropriate, the State) may impose alternate, less stringent, limitations.

Final effluent guidelines and limitations for steam electric generating stations were promulgated by the EPA on 8 October 1974 (40 CFR 423, V. 39 Fed. Reg. 36185-36211). Based upon their evaluation the EPA selected closed cycle cooling (cooling towers) as the best technology. They did, however, exempt from this technology requirement "old units" with once-through cooling systems. They defined an "old unit" as any generating unit which was first placed in service on or before 1 January 1970, and any generating unit of less than 500 megawatts rated net generating capacity which was first placed in service on, or before, 1 January 1974.

Section 316(b) of the Act requires "that the location, design, construction, and capacity of cooling water intake structures reflect the best technology available for minimizing adverse environmental impact." Section 316(b), therefore, specifically addresses the impact of entrainment and entrapment-impingement associated with cooling water systems. Final regulations [40 CFR Part 401 and 402, Section 316(b)] were promulgated on April 26, 1976. A summary of the available technology is provided in the Section 316(b) Development Document (U.S. Environmental Protection Agency, 1976).

Section 316(b) demonstrations are developed in accordance with the appropriate EPA regional or state guidelines, and are handled on a case-by-case basis. An initial assessment is made as to whether the intake has a low or high probable impact. Intakes with low probable impact are defined as those that: (1) withdraw a small fraction of the available water in the contributing water body, (2) are located in biologically unproductive areas, (3) have been shown through historical data to have no adverse effects, or (4) have been characterized by other considerations to have low and acceptable impacts. Intakes with high potential impact usually require extensive documentation to define potential total water body effects. New intakes are provisionally considered to have high potential impact until data are presented to support an alternate conclusion. Section 316(b) studies may be directed at assessing the impact of losses of organisms by cooling water intakes on the populations of organisms in the contributing water body. The EPA (1977) has prepared a guidance manual for assessing the impacts of cooling water intake structures.

V. ASSESSING THE SIGNIFICANCE OF ENTRAINMENT LOSSES

Species composition and relative abundances of entrainable organisms may vary greatly from one plant site to another, and with time at a given location. At a particular site, the potential for pump and plume entrainment depends to a large extent upon the design and placement of the intake and discharge structures.

Before reliable predictions can be made of the likelihood of entrainment problems at a proposed power plant site, and before intake and discharge structure designs can be formulated to minimize entrainment losses, one must have adequate information on the temporal and spatial distributions of the susceptible organisms in the receiving waters. Usually, ichthyoplankton, juvenile fish, and macroinvertebrates are the organisms of primary concern because they have relatively long generation times and are of greater economic value. At some locations phytoplankton, zooplankton, benthic infauna, and other kinds of organisms may be important.

Knowledge of an organism's life cycle, its distribution in the waters contiguous to a plant, and the circulation of those waters are prerequisites to assessing the potential for entrainment. An organism may spend only a portion of its life cycle in the planktonic phase and be susceptible to entrainment for a relatively short, but critical, period. Some fish eggs, winter flounder, for example, are demersal and are not readily subject to entrainment but the larvae are planktonic and may be entrained. During later life stages as larger juveniles and adults they are not entrainable.

In assessing the effects of entrainment, the historic emphasis has been placed on determining the fraction of entrained organisms killed by passage through the plant, and on thermal criteria for minimizing high temperature damage. In the absence of supporting data for physical effects, 100% mortality has frequently been assumed in subsequent analyses of population impacts. Recent studies have indicated that there is a wide range in mortalities of entrained organisms (1% to 100%) from plant to plant, and with species at a particular plant. To make an unequivocal assessment of entrainment effects, the exposed organisms must be held long enough to observe latent effects.

Computer modeling techniques have recently become popular as a method for predicting the effects of entrainment losses on populations. Such population analysis has tackled the difficult task of evaluating the effects of these losses on the local population of organisms, sometimes including subsequent community effects. Although refinements are being made rapidly, modeling has not solved the intractable problems of estimating natural mortality rates and incorporating realistic compensatory mechanisms. Thus, unequivocal conclusions have been elusive.

The significance of entrainment-induced mortality depends upon a variety of factors. The longer the regeneration time, the more susceptible a population of organisms is to persistent effects of a given cropping rate. Regeneration time varies with species, environmental conditions, and productivity of the system. Phytoplankton can regenerate on time scales of hours. Some zooplankton, for example copepods, reproduce in days; others in weeks. Finfish and shellfish characteristically have generation times of from one to several years. Losses of phytoplankton by entrainment rarely, if ever, can be documented outside of the immediate vicinity of a plant and are of little consequence. However, appreciable reductions of zooplankton, for example copepods and opossum shrimp, could occur if several power plants were located in a restricted area. These reductions might be significant if they occurred in an important nursery area for larval and post-larval fishes. These are active and rapidly growing life stages with high food (energy) requirements. If food is insufficient, mortality increases and the survivors may have a longer planktonic existence with prolonged susceptibility to predation. These effects could be manifested in a substantial reduction of a particular year class. Mortality of entrained organisms may represent a double loss to the ecosystem--a loss of the reproductive stages of higher forms and a loss of food organisms for these species (Beck and Miller, 1974), Fig. 3.

Destruction of meroplankton (ichthyoplankton, and planktonic invertebrate larvae)--the reproductive material of higher, commercially important, forms--is the first order of concern.

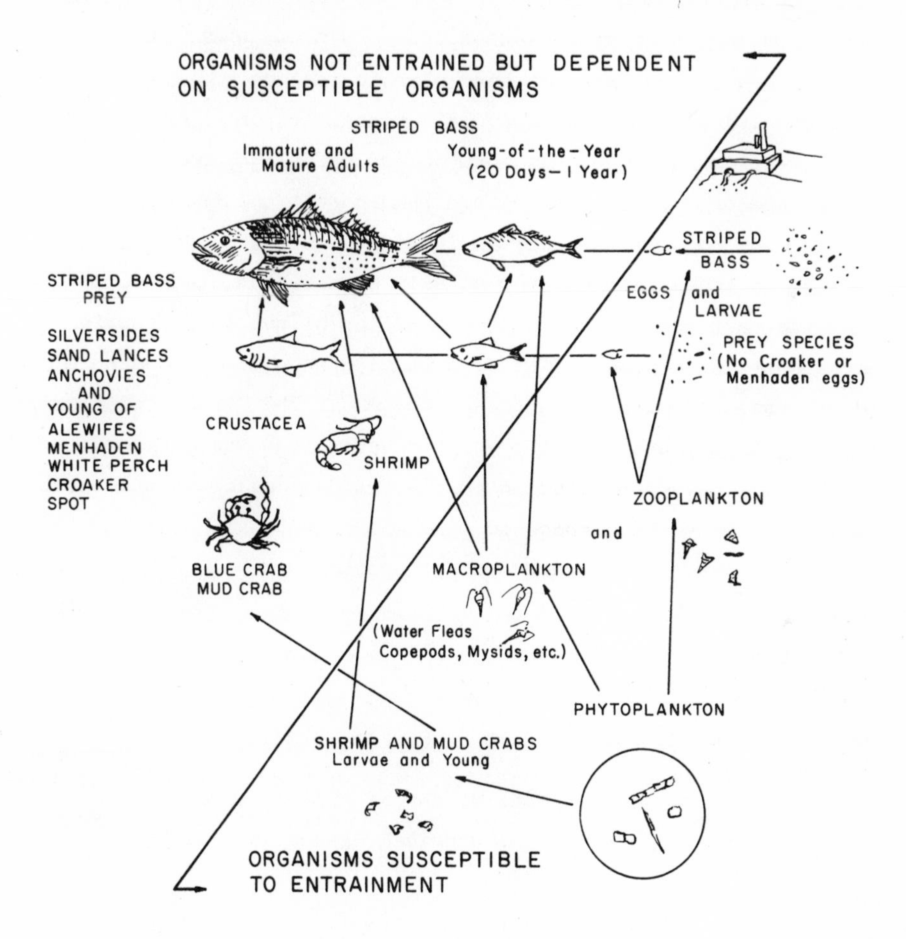

Fig. 3. Susceptibility of striped bass and food web organisms to passage through power plant cooling systems.

This may be a particularly important problem in estuaries and has serious implications for many commercially and recreationally important fishes. McHugh (1966) estimated that nearly two-thirds, by landed value, of the U.S. Atlantic commercial catch of fishes and invertebrates are estuarine-dependent species. Meroplankton is a major component of the estuarine ecosystem, where about 70% of the fauna is estimated to have planktonic eggs and larvae (Thorson, 1946, 1950). Unequivocal assessments of the effects of entrainment losses on populations of ecologically important species and on community structure remain critical problems.

Because population analysis is no panacea for judging whether entrainment losses are environmentally significant, this committee feels that a return to careful examination of the damage during entrainment is in order. This examination should yield methods and criteria for minimizing these damages at the "front end," thus simplifying judgments about population effects. The extraordinarily high costs of gathering population data at certain notable power plant sites has suggested that minimizing damages in the first place may be much more cost-effective.

VI. GOALS OF THE REPORT

The goals of this report are:

(1) to assess the effects of the several stresses--thermal, physical and chemical--associated with entrainment.

(2) to use the results of this assessment to develop guidelines for the conceptual design and operation of power plants with once-through cooling systems to minimize the mortalities of entrained organisms, and to minimize the effects of power plant cropping on the populations of those organisms.

(3) to outline research priorities required:

(a) for a significant improvement in our ability to make an unequivocal assessment of the mortalities associated with entrainment;

(b) to resolve the effects of each of the several stresses; and

(c) to improve the design and operating criteria of power plants with open-cycle cooling systems to ensure acceptable and predictable mortalities due to entrainment.

REFERENCES

Beck, A. D. and D. C. Miller. 1974. Analysis of inner plant passage of estuarine biota. Pages 199-226 *in* Proc. Amer. Soc. Civ. Eng. Power Div. Spec. Conf., Boulder, Colo., 12-14 August 1974.

Calvert Cliffs "Coordinating Committee" vs. Atomic Energy Commission. 449 F. 2 d 1109, cert. denied 404 U.S. 492 (1972).

Federal Water Pollution Control Act. Amendments of 1972, Public Law 92-500, October 18, 1972 (33 U.P.S.C. 1151).

McHugh, J. L. 1966. Management of estuarine fisheries. Pages 133-154 *in* A Symposium on Estuarine Fisheries. Am. Fish. Soc. Spec. Publ. 3.

National Environmental Policy Act of 1969, Public Law 91-190, January 1, 1970. [42-U.P.S.C. 4321-4347; see paragraph 4332(c)].

National Pollutant Discharge Elimination System, Section 402 of the FWPCA (PL92-500) October 18, 1972.

Thorson, G. 1946. Reproductive and larval development of Danish marine bottom invertebrates--with special reference to planktonic larvae in the Sound, Cøresund. Meddeleser Komm. Danmarks Fiskeri Havundersogeleser Ser. Plankton 4. 523 p.

Thorson, G. 1950. Reproduction and larval ecology of marine bottom invertebrates. Cambridge Philosophical Society Biological Reviews 25: 1-45.

U.S. Atomic Energy Commission. 1969. Thermal effects studies by nuclear power plant licenses and applicants. U.S. Government Printing Office, Washington, D.C.

U.S. Environmental Protection Agency. 1976. Development document for best technology available for the location, design, construction and capacity of cooling water intake structures for minimizing adverse environmental impact. EPA 440/1-76/015-a. Washington, D.C. 263 p.

U.S. Environmental Protection Agency. 1977. Guidance for evaluating the adverse impact of cooling water intake structures on the aquatic environment: Section 316(b) P.L.92-500, A Draft. U.S. Environmental Protection Agency, Office of Water Enforcement, Permits Division, Industrial Permits Branch, Washington, D.C., May 1, 1977. 59 p.

Van Winkle, W., ed. 1977. Proceedings of the Conference on Assessing the Effects of Power-Plant-Induced Mortality on Fish Populations, Gatlinburg, Tennessee, May 3-6, 1977. Pergamon Press, N.Y. 439 p.

CHAPTER 2. THERMAL EFFECTS OF ENTRAINMENT

J. R. SCHUBEL

C. C. COUTANT

P. M. J. WOODHEAD

TABLE OF CONTENTS

I. INTRODUCTION

Organisms may be drawn into the cooling water intake of a power plant, passed through the condensers and discharged back into the environment; or they may be entrained into the discharge plume along with the diluting water without having passed through the plant. The organisms that pass through the plant are subjected initially to a very rapid rise in temperature, approximately equivalent to the temperature rise across the condensers. They are exposed to this maximum excess temperature (the temperature above that which would be observed if the plant were not generating, and commonly designated by ΔT) during passage through the plant and to the point of discharge, and then to decreasing excess temperatures as they are carried down the plume. In the near field, the rate of cooling along the plume is determined almost entirely by the intensity of the lateral and vertical mixing. Organisms that are entrained into the discharge plume without having passed through the plant are subjected to a less rapid rise in temperature and to somewhat lower excess temperatures. The actual time-excess temperature histories experienced by entrained organisms vary from plant to plant, and to a much smaller extent, with time at any given plant. The time-excess temperature exposure history depends upon the design of a plant's cooling water system, particularly its discharge structure, the plant's operating load, the flow rate of cooling water, and the characteristics of the receiving waters.

The heat exchange in the plant takes place in the condenser units. Each condenser unit consists of a bank of up to 10,000 tubes, each approximately 2.5 cm in diameter and from 10 to 25 m long. Plants may have from one to six condenser units. The cooling water passes through the inside of the condenser tubes while steam from the turbine exhaust passes around the outside. Since the flow velocity of the cooling water through the condensers is usually about 1.5 to 6 m/sec, heat is added to the cooling water for only 2 to 20 sec. The rise in the temperature of the cooling water is therefore almost instantaneous.

The cooling water of a steam electric plant may be discharged over a weir, through a canal, a multi-port diffuser, or as a high speed surface or submerged jet. The most rapid dilution of the heated effluent, and therefore the most rapid reduction in temperature, is attained with long diffusers that discharge the water through a large number of small ports--multi-port diffusers. Submerged jet discharges are the next most effective mode of discharge in terms of temperature reduction, and are considerably less expensive to construct than multi-port diffusers. Canals are the least desirable type of discharge structure from the standpoint of thermal effects because relatively little cooling takes place within a canal unless colder diluting water is added with auxiliary pumps.

The extent of the region of excess temperature and the distribution of excess temperature in the waters contiguous to a plant are determined by the volume of cooling water pumped, the fraction of the available water used for cooling, the size and design of the plant, the design and placement of the discharge structure, and the hydrodynamic characteristics of the receiving waters.

It has long been known that abrupt thermal shocks may disturb normal processes in the development of early life stages of aquatic organisms, or result in death of either young or adults (Kinne, 1970). Organisms which are entrained and passed through a power plant's cooling system are subjected to a sequence of thermal stresses characteristic of a particular plant. The thermal stresses experienced during passage through the cooling system, and subsequently in the discharge waters, may result in physiological damage, debilitation, and/or death. The degree of thermal damage depends upon the increment (ΔT) above ambient water temperature, upon the absolute value of the maximum temperature, and upon the duration of the thermal stress; longer exposures have greater potential for damage. This chapter addresses the thermal effects experienced by entrained organisms.

The objectives of this chapter are:

(1) to describe the range of time-excess temperature histories typical of operating and proposed steam electric stations;

(2) to describe a conceptual framework for predicting thermal effects on organisms entrained by steam electric plants with once-through cooling systems;

(3) to critically review the literature on thermal tolerances of a variety of aquatic organisms and to assess whether these data can be used in this conceptual framework;

(4) to assess the usefulness of field studies of entrainment in estimating thermally-induced mortalities;

(5) to outline the types of studies required both for a significant improvement in our ability to predict and assess the thermal effects of entrainment, and

(6) to specify, on a conceptual basis, design and operating criteria for power plants with once-through cooling systems to minimize entrainment losses caused by *thermal* stress.

II. SOME TYPICAL TIME-EXCESS TEMPERATURE HISTORIES

Adams (1969) presented a detailed time-excess temperature history for passage through the Pittsburgh power plant of the Pacific Gas and Electric Company. Pritchard and Carter (1972) developed, for a variety of power plant design and operating criteria, time-excess temperature relationships that give the maximum time of exposure of entrained organisms to any specified ΔT. Each relationship is based on a specific maximum ΔT, corresponding to the temperature rise across the condensers, and on a particular set of cooling conditions; conditions ranging from surface cooling only, to surface cooling with various degrees of dilution. The rate of dilution is dependent on a variety of factors including: the speed of the discharge at the discharge orifice, the width of the orifice, and the circulation of the receiving waters. Using data from Pritchard and Carter (1972), Schubel (1974) presented a number of time-excess temperature histories characteristic of a variety of power plant designs.

Coutant's (1970b) hypothetical time-excess temperature curves for a power plant with a multi-port diffuser discharge and for a plant with a discharge canal are reproduced in Fig. 1. A time-excess temperature history typical of a plant with a jet discharge is presented in Fig. 2.

There have been several summaries of maximum ΔT's. The Water Resources Council (1968) reported that the average rise across the condensers of the U.S. power plants--both fossil fuel and nuclear--was $8.3^{\circ}C$. According to Coutant's (1970b) examination of 61 nuclear power plant designs filed with the U.S. Atomic Energy Commission, the predicted ΔT for these plants was $10.8^{\circ}C$, and the range from 5.6 to $18.6^{\circ}C$. A more recent examination of the maximum ΔT's of operating and proposed nuclear power plants

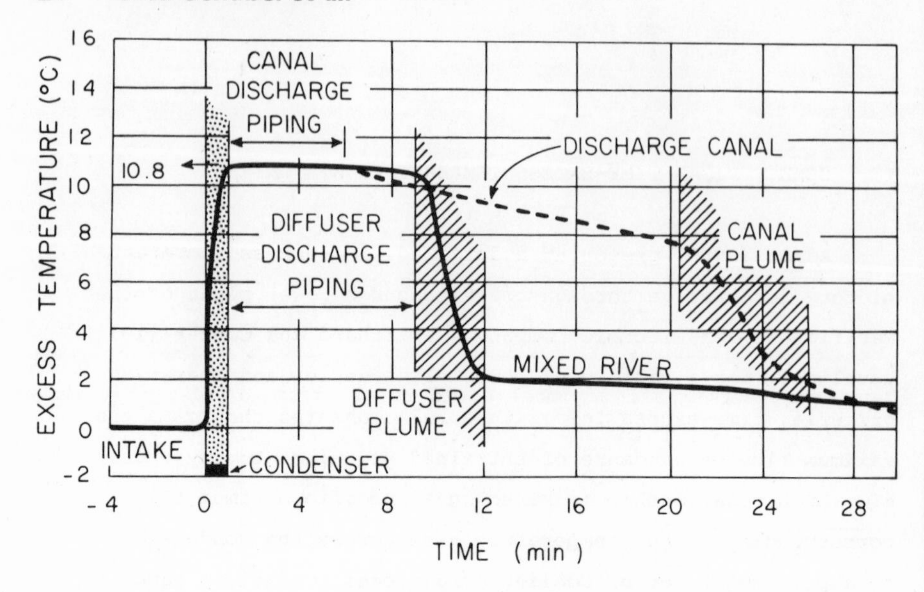

Fig. 1. Hypothetical time-courses of acute thermal shock to organisms entrained in condenser cooling water and discharged by diffuser or via a discharge canal (after Coutant, 1970b).

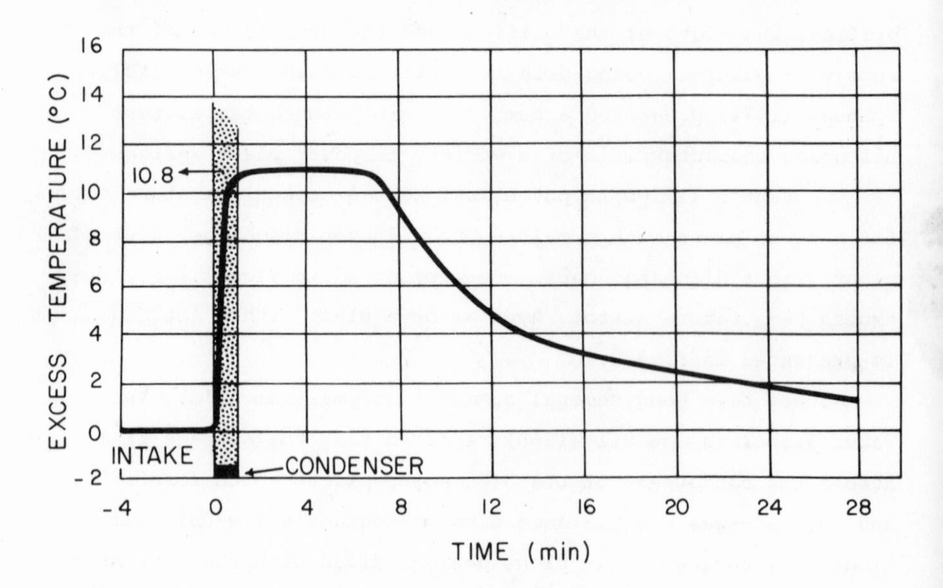

Fig. 2. Hypothetical time-course of acute thermal shock to organisms entrained in condenser cooling water and discharged as a jet.

indicated a mean (weighted by number of plants) temperature rise across the condensers of 11°C, and a range of from less than 6 to more than 19°C (Fig. 2, Chapter 1).

Excess temperatures much less than 7.7°C are, according to Lee (1970), generally impractical because of the economics of pumping costs and equipment size. However, some states, Maryland for example, limit the ΔT at new plants to only 5.6°C. This stringent criterion was presumably adopted because of anticipated adverse thermal effects that could result from the use of a higher ΔT.

III. A CONCEPTUAL FRAMEWORK FOR PREDICTING THERMAL EFFECTS OF ENTRAINMENT

There is a well-established conceptual framework for predicting thermal damages to aquatic organisms entrained by power plants with once-through cooling systems (Coutant, 1970a; 1972). This framework is based on a relatively large number of controlled laboratory experiments with a variety of kinds and sizes of organisms--mostly adults--and on a smaller number of field studies at operating power plants. The conceptual framework is sound, but reliable predictions of the thermal effects of entrainment are possible only for those species for which appropriate thermal tolerance data are available. As we shall see, these cases are relatively rare because field data are usually equivocal, and because most laboratory experiments have not been designed to assess the thermal effects of relatively short exposures to rapidly varying temperatures.

This section summarizes methods of analysis of thermal tolerance data that can be used to determine whether or not organisms will be capable of surviving the thermal stresses associated with entrainment. At this point, no consideration is given to potential interactions of thermal stresses with physical or chemical stresses which may occur simultaneously.

A. Upper Incipient Lethal Temperature and Dose Response for Thermal Death

It has been known since the 1940's that fish[1] have a discrete temperature tolerance range which varies with acclimation temperature and is characteristic for each species. Temperatures which exceed this range induce mortality which is time-dependent; the higher the temperature above the tolerance limit, the more rapid is death (see, for example, reviews by Fry et al., 1946, 1971, and Brett, 1952, 1960). The ends of this range, or lethal thresholds, are usually defined as 50% survival of a sample of fish. Lethal thresholds typically are referred to as "incipient lethal temperatures." The time-to-death response which follows the intensity of the thermal stressor is conceptually analogous to dose-response patterns in chemical toxicity, action of pharmaceuticals, and radiation damages. Thus, for each exposure temperature above an upper tolerance threshold (incipient lethal temperature), a finite survival time can be ascribed, Fig. 3, although this tends to vary somewhat according to a host of environmental factors such as day length, time-of-day, salinity or hardness of the water, and the physiological state of the organisms (Hutchinson, 1976). Of particular importance, are differences in developmental stage, age and size--all of which are obviously related. For some organisms, the temperature range over which survival is time dependent is so small that the lethal threshold alone is sufficient to define survival temperature requirements.

As usually presented in the literature, thermal resistance data do not provide a sufficiently complete description of survival for assessing the importance of entrainment losses. It

[1] *This was later confirmed for other organisms.*

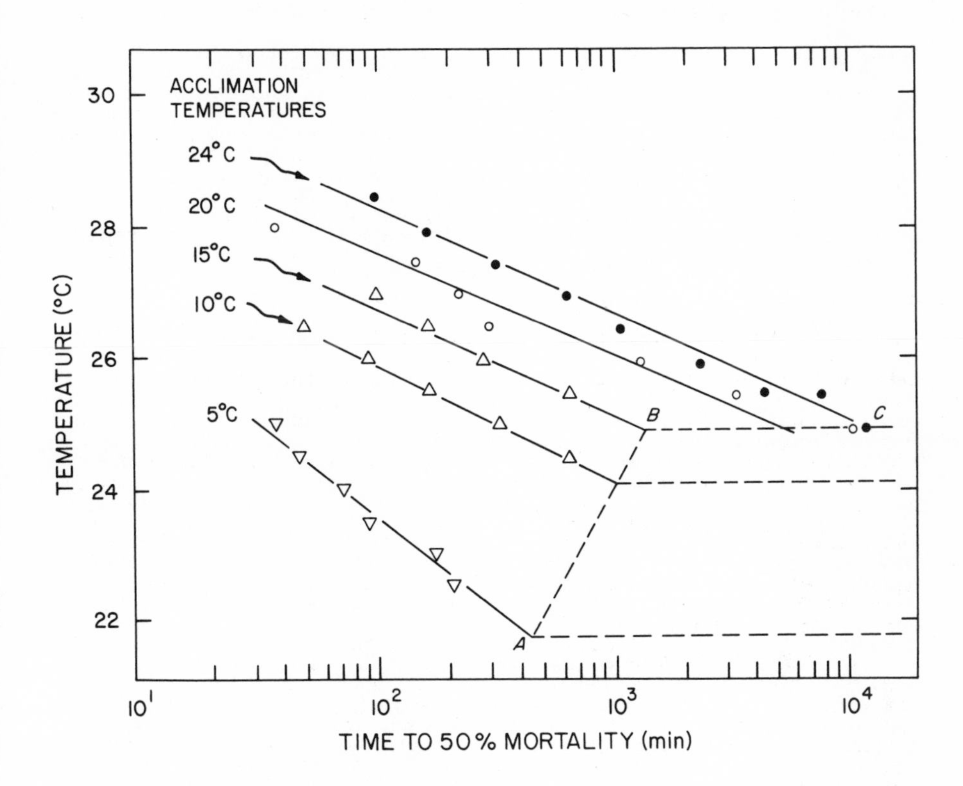

Fig. 3. Median resistance times to high temperature of organisms acclimated to different temperatures. For each acclimation temperature there is an incipient lethal temperature--the highest temperature to which an organism can be continuously exposed for an indefinite period without increasing the mortality rate. Line A-B denotes rising lethal threshold temperatures with increasing acclimation temperature. This rise in threshold eventually ceases at B-C, the ultimate incipient lethal temperature. *(Adapted from Brett, 1952).*

may be important, for example, to know what thermal dose will cause mortality rates other than 50% (Hoss et al., 1974). On the basis of population dynamics, a loss of up to 30% of the larvae of a species may be acceptable and we may wish to know the time-temperature limitations for this endpoint. We could improve the usefulness of thermal resistance curves such as those plotted in Fig. 3, if each of the regression lines were replaced by a family of curves representing mortality rates ranging from 10 to 90% at 10% intervals. Kennedy et al. (1974a) presented 10 and 90% mortality data for their study of the thermal resistance of hard clam (*Mercenaria mercenaria*) larvae. Another deficiency of thermal resistance data from "conventional" experiments, is the lack of a rapid rate of cooling at the end of the exposure period; cooling that is characteristic of entrainment and which may be an important source of damage (Hoss et al., 1974). The most appropriate method for determining thermal effects for application to entrainment problems may be through a series of discrete, square-wave, time-temperature exposures for which percentages of mortality are determined.

Another form of high and low temperature tolerance information, the "critical thermal maximum" (or minimum) has recently gained popularity among thermal researchers (Hutchinson, 1961; Gibbons and Sharitz, 1974; Esch and McFarlane, 1976). This trend is regrettable because thermal tolerance information in the form of Critical Thermal Maxima (CTM) can not be used as a predictive tool for entrainment and is useful mostly for screening organisms for relative thermal sensitivity (Fry, 1967; Coutant, 1970; Hair, 1971; Hoss et al., 1974). The CTM is based on an ecologically plausible concept that loss of equilibrium is an important endpoint for ecological survival, and that during temperature rise or fall a temperature endpoint can be observed "at which locomotory activity becomes disorganized and the animal loses its ability to escape from conditions that will promptly lead to its death" (Cowles and Bogert, 1944).

It is not the concept of equilibrium loss that is at fault, but the way in which the endpoint is determined in the CTM methodology. This methodology mixes the two operating variables--temperature and time of exposure--in a manner that is neither standardized nor amenable to separation of the variables. The CTM has been determined by heating animals at a constant rate that is just fast enough to allow the deep body temperature to track the ambient test temperature (Hutchinson, 1976). This implies different rates of heating according to body size, circulation, and thermal conductivity. Different investigators have used widely varying heating rates but without regard to body sizes; heating rates of $1^{\circ}C$ per min and $1^{\circ}C$ per hr are common. None of these rates simulates temperature changes experienced during entrainment, nor does the merging of variables allow translation of the data to real entrainment exposures.

The classical 24-, 48-, and 96-hr exposure times to calculate LT_{50} are equally inappropriate to power plant entrainment assessments. In these procedures, samples of organisms are placed in water baths at each of several constant temperatures, and the percentage of deaths is recorded for each batch at the end of the given time periods. A resistance pattern can be obtained that identifies both time and temperature variables, but not in the range of exposure times (usually a few min to a few hrs at most) characteristic of entrainment. Data are needed at shorter exposure times. Burton et al. (1976) describe an example of how one could draw an incorrect conclusion from LT_{50} data. The 48 hr LT_{50} for mysid (*Neomysis awatschensis*) is $22.5^{\circ}C$ when acclimated to $11^{\circ}C$, and $25^{\circ}C$ when acclimated to $22^{\circ}C$ (Hair, 1971). But, Burton et al. (1976) observed no mortality of this mysid following a 6 min entrainment exposure to a ΔT of $25^{\circ}C$ above a base temperature of $10^{\circ}C$. Another example was the study of Lauer et al. (1974) in which it was concluded that for most species in their Hudson River study, use of 24-, 48-, and 96-hr tolerance data would have led to predictions of 100% mortalities, yet, in fact, their bioassay results for the exposure times (5-60 min) characteristic of the Indian Point plant, New York, as well as field observations at that plant, indicated that nearly 100% of the entrained organisms survived.

Entrainment at power plants involves egg, larval, and young juvenile stages of fishes as well as planktonic invertebrates, phytoplankton, and zooplankton. Since most of the thermal resistance data are for large fish, they are of use principally for conceptual guidance. There are however, sufficient data on small organisms and early life stages to establish the general applicability of the dose-response concept to entrainable organisms. Development of an adequate predictability for survival of planktonic assemblages depends upon amassing time-temperature survival data for the relevant species and life stages. We should devote far more attention to this task than we have in the past.

B. Pre-death Debilitation

Again borrowing concepts from work with large fishes, it is apparent that organisms go through progressive debilitation under high temperature stress prior to actual death, and that this debilitation can have important consequences for an organism's survival in a natural ecosystem. In experiments by Coutant and Dean (1972), heat-stressed salmonids lost equilibrium a considerable time before they died. In subsequent studies Coutant (1973) demonstrated that susceptibility of trout fry to predation was increased at exposure times only 11% of those necessary for death. Significantly, the endpoints for both equilibrium loss and increased predation also followed a "dose response" pattern in which the effect occurred sooner at higher temperatures (Fig. 4) than at lower temperatures, and the effect could be carefully defined by two variables, temperature and time. Other studies of survival in a laboratory predator-prey system showed that sudden temperature increases of 10°C above acclimation (at 7, 12 and 17°C) increased the vulnerability of sockeye salmon

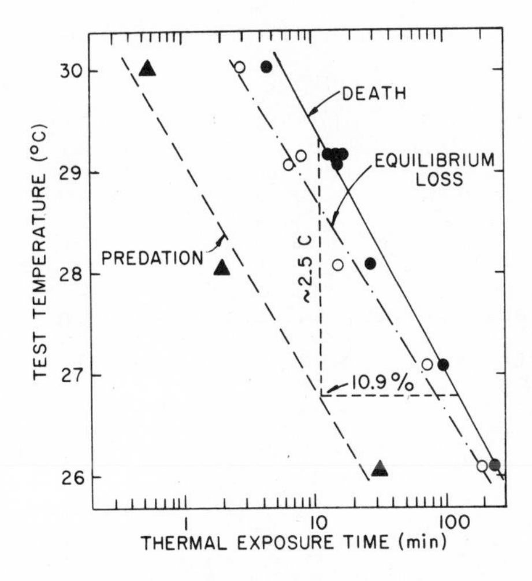

Fig. 4. Relationships among three effects of acute thermal shock on a sibling group of juvenile rainbow trout acclimated to 15°C: (1) time to initial increase in vulnerability to predation, (2) median time to loss of equilibrium, and (3) median death time. (From Coutant, 1973).

fry (*Oncorhynchus nerka*) to predation by yearling coho salmon (*O. kisutch*) (Sylvester, 1971, 1972). Similarly, Yocom and Edsall (1974) acclimated fry of lake whitefish (*Coregonus clupeaformis*) to 5, 10, 15, and 18°C, and exposed them for one minute to ΔT's of 19, 15, 13 and 11°C respectively. They then placed them, and unshocked control fry, in tanks containing larger predators--yellow perch (*Perca flavescens*). The number of predator attacks was recorded along with the number of fry eaten. The brief heat shocks increased the vulnerability of the whitefish fry; more were captured per attack than in the control groups.

If the studies with trout and salmon (Coutant 1973) are representative, and they may not be, debilitation sufficient to significantly increase predation occurs at temperatures about 2°C lower than temperatures necessary for death at the same exposure time. A "correction factor" linking debilitation times with death times can be useful, since much of the literature data are for death, and determinations of mortality are far simpler than are experimental predation studies. A 2°C "correction factor" has been proposed for use with thermal resistance data (National

Academy of Sciences, 1973).

C. Sublethal Effects

Because the duration of entrainment is usually short, little attention has been given to sublethal thermal effects other than reports of disorientation or debilitation. But eggs and larvae are highly sensitive life stages undergoing rapid changes in the processes of cell and tissue determination and differentiation; some stages are much more sensitive to temperature than others (Needham, 1942; Frank, 1974). Temperature elevation at such times might cause the shift of a determinative process and throw it out of coordination with other developmental processes, leading to abnormalities in later life.

Borrowing from studies with larger fish, entrainment would be expected to induce physiological responses to the stresses experienced. A general physiological response to nonspecific stresses, including abrupt temperature changes of 10°C or less, has been well established for fish and is similar to that in higher vertebrates. The response is rapid and has been shown to occur within 2.5 min of the onset of shock (Chavin, 1964). The response consists of a complex of reactions including ionic changes in the blood and tissues, and changes in counts of circulating red corpuscles and especially of white cells, accompanied by an increase in corticosteroid secretion by the interrenals (Reaves et al., 1968; Wedemeyer, 1969; Pickford et al., 1971). The response may be protracted, persisting long after termination of the shock. The reported times for recovery range from 2 hr to a day, or more, in different studies. Corticosteroids, like other hormones, can have profound effects upon morphogenesis in developing larvae, affecting some processes such as bone and cartilage development, more than others. Through the actions of hormones, the protracted physiological response to the brief stresses of entrainment could disturb the normal organization of morphogenesis and cause anomalies in later development. Unfortunately, biological studies of the effects of power plant entrainment have not addressed such subtle sublethal effects. It has been shown that gross deformities may develop in response to thermal shocks lasting only minutes during egg stages (e.g., Bergan, 1960; Hopkins and Dean, 1975). The general literature of subtle sublethal effects in the development of fish eggs and larvae, which can be produced by a wide variety of stressors, both physical and chemical, has been recently reviewed by Rosenthal and Alderdice (1976).

D. Population Responses

Our primary concern for aquatic organisms rests with populations, not individuals as it does in human communities. Therefore, any tests of mortality, whether of direct or "ecological" death, should be viewed in the context of predicting losses to recruitment of individuals to a steady (or otherwise socially or ecologically desirable) population size. At the community or ecosystems levels we should also be concerned with the superposition of power-plant-induced mortality on natural mortality. At this point, however, we can justifiably leave such concerns until we have determined whether the thermal stresses of entrainment would, in fact, cause death to those fractions of the population that are entrained.

E. Using Time-Temperature Data

Having obtained the fundamental time-temperature data for the species and life-stage in question through controlled experiments, how are they applied to entrainment?

The straightforward use of the time-temperature graph (such as Fig. 5) as a nomogram for the power plant's intake and condenser discharge temperatures may often suffice. For example, a power station may withdraw water at 20°C and discharge it

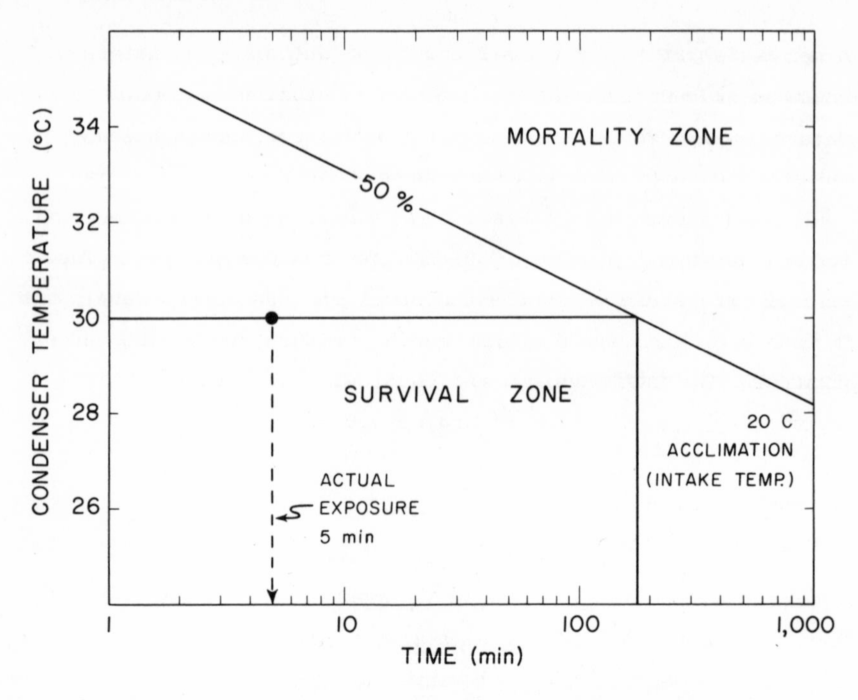

Fig. 5. Example of how a Thermal Resistance Curve can be used to predict whether or not mortality from thermal stresses will be expected for entrained organisms. Consider a plant with an intake temperature of 20°C, a ΔT of 10°C, and a transit time of 5 min. Assume the organisms are acclimated to 20°C and that cooling takes place instantaneously.

through a rapid diffuser at 30°C where mixing with ambient temperature water reduces the effluent to essentially 20°C almost instantaneously; travel time from condenser to diffuser has been measured (with dyes, for example) and is about 5 min. This exposure pattern is compared directly with the time-temperature graph for survival for the species to be protected which may indicate that a 20°C-acclimated representative of this species (50% mortality data) will survive at 30°C for 3 hr. Data may also be available which indicate 10% of the sample tested died after an exposure of 2.5 hr and 90% died after 3.5 hr. Thus, the thermal exposure BY ITSELF would not be expected to cause mortality.

A data set for equilibrium loss or susceptibility to predation could have been used for the species if it were available, or the data set for 50% mortality could have been adjusted downward by about $2^{\circ}C$ to estimate a predation threshold.

For convenience in summarizing large amounts of time-temperature data, tables of coefficients for semilogarithmic regression equations have replaced graphs (National Academy of Sciences, 1973), but the comparisons remain equally straightforward. The basic equation,

$$\log TIME_{(min)} = a + b\,(TEMP_{(^{\circ}C)}) \tag{1}$$

is used to calculate corresponding times and temperatures for 50% mortality. By rearrangement, the equation can also be used to define conditions for survival by setting the right side of the equation to less than or equal to 1. This gives:

$$1 \geq \frac{TIME_{(min)}}{10^{a + b\,(TEMP_{(^{\circ}C)})}} \tag{2}$$

and by incorporating the $2^{\circ}C$ "correction factor" for debilitory effects prior to death we have:

$$1 \geq \frac{TIME_{(min)}}{10^{a + b\,(TEMP_{(^{\circ}C)} + 2_{(^{\circ}C)})}} \tag{3}$$

Thermal exposures during entrainment are generally not as simple as just described, however. While heating is generally abrupt, and a portion of the exposure is at reasonably constant temperatures in pipes and conduits, the rate of cooling varies considerably with discharge structure (Figs. 1 and 2). To evaluate such exposures, one must rely upon the additivity of thermal damages first explored by Fry, Hart and Walker (1946) and subsequently confirmed by others.

Exposures to changing temperatures during entrainment can be viewed as a sequence of discrete temperature exposures each having a known duration (Fig. 6). Rearrangement of the basic semi-

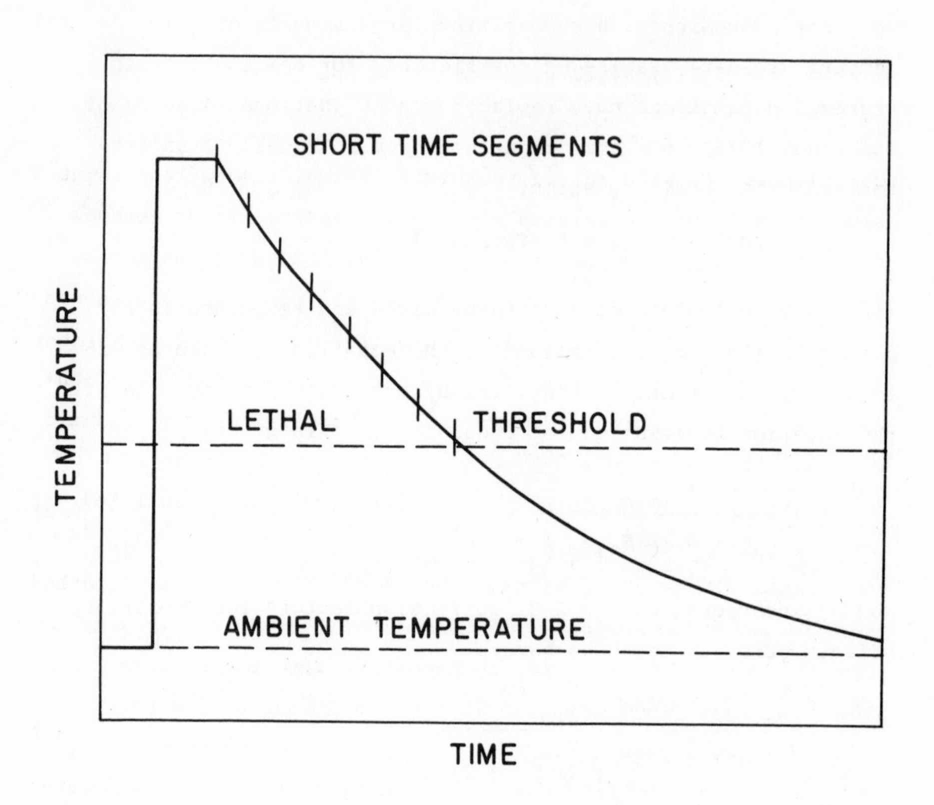

Fig. 6. Representative time-temperature exposure history experienced during entrainment broken down into short-time segments for calculation of additive thermal doses.

logarithmic equation (Eq. (1)) in the form of Eq. (2) or Eq. (3) is particularly useful now, for the increments can be added until the temperature falls below the lethal threshold, thus:

$$1 \geq \frac{TIME_1}{10^{[a + b \ \ TEMP_1)]}} + \frac{TIME_2}{10^{[a + b \ \ TEMP_2)]}} + \cdot \cdot \cdot + \frac{TIME_n}{10^{[a + b \ (TEMP_n)]}} \tag{4}$$

The computation is readily handled by a programmable desk calculator. As with Eq.(2), survival is indicated if the right side of the equation remains at or below unity. These methods have been introduced previously (National Academy of Sciences, 1973) and they have been used in Environmental Impact Statements for the U.S. Atomic Energy Commission (now U.S. Nuclear Regulatory Commission). A recent compilation of time-temperature relationships and lethal threshold temperatures for a variety of aquatic organisms but particularly fish, is contained in Appendix II-C of the Environmental Protection Agency's 1973 report entitled "Water Quality Criteria 1972."

It should be apparent that these data and calculation methods can be used as design criteria as well as methods to evaluate already selected designs. With the known time-temperature responses of key species as boundary conditions to ensure survival, the engineer can select pumps, discharge structures, and mixing zones to find a safe combination of ΔT's and exposure times that will suit his particular plant site.

At the present state of power plant entrainment analysis, the foregoing can be considered as a hypothesis which is ripe for critical appraisal against the mounting body of experimental and field evidence. The conceptual framework should also be used in the design of laboratory experiments intended to predict the thermal effects of entrainment and to provide guidance in the selection of power plant design and operating criteria to minimize thermal effects.

IV. A BRIEF REVIEW OF THE LITERATURE

A. Introduction

The purpose of this section is to assess the effects of the thermal stresses experienced by zooplankton, macroinvertebrates, ichthyoplankton, and juvenile fishes during entrainment by power plants with once-through cooling systems with a variety of design and operating criteria. The assessment is based on laboratory studies in which appropriate time-temperature histories have been used and on field studies, both published and unpublished. We have not considered phytoplankton. Regeneration times of phytoplankton are so short relative to those of zooplankton, and particularly ichthyoplankton, that protection of these latter groups of organisms should ensure adequate protection of phytoplankton.

Scientists have attempted to assess the thermal effects of entrainment by conducting extensive and expensive site studies at many operating power plants, by examining the classical biological literature on thermal effects, and more recently by subjecting organisms in the laboratory to time-excess temperature histories representative of those experienced by organisms entrained by power plants. We shall assess what we have learned from these approaches which is useful in predicting the thermal effects of entrainment.

Although there is a vast literature on the effects of temperature on a wide variety of aquatic organisms (see, for example, the bibliographies of Coutant, 1968, 1969, 1970a, 1971a; Coutant and Goodyear, 1972; Coutant and Pfuderer, 1973, 1974; Coutant and Talmage, 1975, 1976; Kennedy and Mihursky, 1967; Raney et al., 1973; Beltz et al., 1974), relatively little of this research is of *direct* use in predicting the effects of exposure to time-temperature histories typically experienced during entrainment by power plants with once-through cooling systems. In nearly all of the published laboratory thermal studies, the organisms have been subjected to a constant temperature for all, or nearly all, of the experimental periods. As we have seen, such exposure histories are clearly not representative of those experienced by organisms carried through the once-through cooling systems of steam electric plants.

B. Ichthyoplankton

1. *Eggs*

a. Laboratory studies. In the past several years an increasing number of investigators have exposed eggs and larvae of a variety of fishes in the laboratory to time-excess temperature histories typical of those experienced by organisms entrained by power plants. The first reported experiments with fish eggs that simulated the thermal experiences of entrained organisms were those of Schubel and Auld (1972a, 1972b, 1973, 1974).

Schubel and Auld (1972a,b; 1973, 1974), Schubel (1974), Schubel and Koo (1976) and Schubel et al. (1976) exposed eggs of blueback herring (*Alosa aestivalis*), American shad (*Alosa sapidissima*), alewife (*Alosa pseudoharengus*), white perch (*Morone americana*) and striped bass (*Morone saxatilis*) to time-excess temperature histories typical of power plants with once-through cooling systems with a variety of design and operating features. These investigators used ΔT's ranging from about $5^{\circ}C$ to $20^{\circ}C$, and in every case superimposed the ΔT's on the average surface water temperature on the spawning grounds when the particular ripe female was caught.

The base temperatures used in Schubel's (1974) experiments are summarized in Table 1. Their set of time-excess temperature exposure curves are contained within the envelope shown in Fig. 7.

Schubel and Auld (1972a,b; 1973; 1974) and Schubel (1974) reported that exposure of eggs of these species to ΔT's of up to $10^{\circ}C$ with the time-excess temperature histories shown in Fig. 7

TABLE 1 Summary of Base Temperatures Used in Schubel's (1974) Experiments

Species	*Range in Base Temperature °C*	*Mean Base Temperature °C*
Blueback herring (Alosa aestivalis)	*15.0 - 18.3*	*17.6*
Alewife (Alosa pseudoharengus)	*12.0 - 14.5*	*13.0*
American shad (Alosa sapidissima)	*16.5*	*16.5*
White perch (Morone americana)	*13.5 - 14.5*	*14.0*
Striped bass (Morone saxatilis)	*14.5 - 18.3*	*17.0*

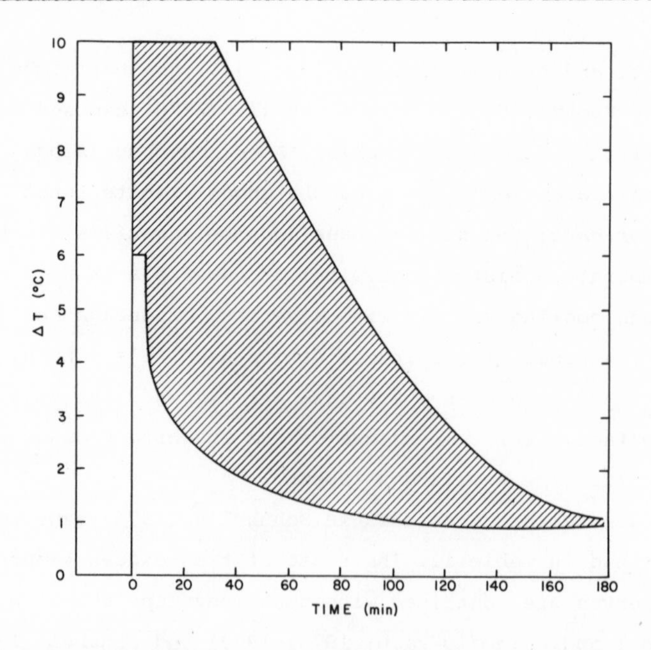

Fig. 7. The envelope of time-temperature exposure histories used in experiments by Schubel and Auld (1972a,b; 1973; 1974) and Schubel (1974).

did not significantly affect survival, hatching success, or morphological development. Eggs in a variety of stages of development were tested.

Schubel and Koo (1976) and Schubel et al. (1976) extended these experiments to higher ΔT's--15 and $20^{\circ}C$--for blueback herring, American shad and striped bass eggs and larvae. The ΔT's were again keyed to the average temperatures on the spawning grounds, Table 2. The range of time-excess temperature curves

TABLE 2 Summary of Base Temperatures Used by Schubel and Koo (1976) and Schubel et al. (1976)

Species	*Range in Base Temperature °C*	*Mean Base Temperature °C*
Blueback herring (Alosa aestivalis)	*17.9 - 21.1*	*19.6*
American shad (Alosa sapidissima)	*20.2 - 20.5*	*20.5*
Striped bass (Morone saxatilis)	*16.6 - 19.6*	*18.5*

were contained within the envelope shown in Fig. 8. An excess temperature of $20^{\circ}C$ resulted in nearly total mortality for eggs of all three species for all exposure histories. At a ΔT of $15^{\circ}C$ all exposure histories significantly reduced the hatching success of both blueback herring and American shad eggs, but none of the exposure histories significantly reduced the hatching success of striped bass eggs.

Schubel et al. (1976) did not report differences in thermal sensitivity of eggs in different stages of development, but they tested over only a limited range of development--early to late embryo.

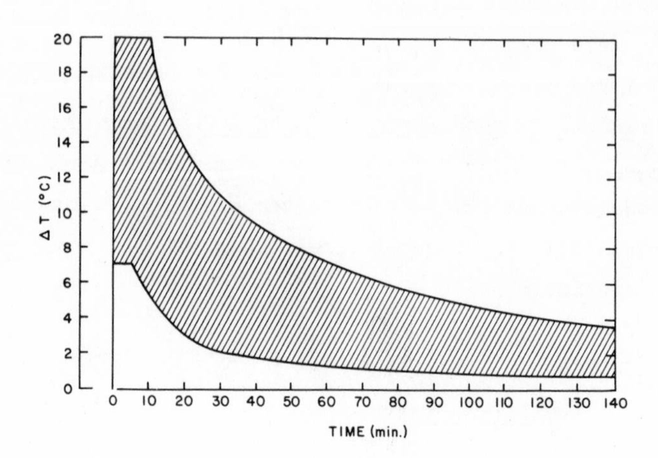

Fig. 8. The envelope of time-temperature exposure histories used by Schubel and Koo (1976) and Schubel et al. (1976).

Koo et al. (in prep.) continued the experiments of Schubel and colleagues. They subjected alewife, American shad, and striped bass eggs to ΔT's of 10.0, 11.5, 13.0 and 14.5°C using time-excess temperature exposure histories similar to those reported by Schubel (1974) and Schubel et al. (1976). The base temperatures used by Koo et al. (in prep.) are summarized in Table 3. The base temperature for American shad and striped bass are higher than those used for the same species by Schubel (1974) and Schubel et al. (1976).

Koo et al. (in prep.) reported that at base temperatures of up to 14.6°C alewife eggs could withstand exposure to a maximum ΔT of at least 14.5°C for up to 5 min without a significant reduction in hatching success. The period of cooling extended over three hours. At a base temperature of 13.9°C exposure to a maximum ΔT of 14.5°C for 15 min significantly reduced hatching

TABLE 3 *Summary of Base Temperatures Used by Koo et al. (in prep.)*

Species	*Range of Base Temperature °C*	*Mean Base Temperature °C*
Alewife (Alosa pseudoharengus)	12.5 - 14.6	13.9
American shad (Alosa sapidissima)	17.0 - 24.0	18.8
Striped bass (Morone saxatilis)	20.5 - 26.5	23.9

success in four experiments. Exposure of eggs to the same time-excess temperature history of eggs from four different females and acclimated to a slightly higher base temperature of 14.5°C did not significantly reduce hatching success. The apparent discrepancy was attributed to variations in thermal sensitivity of eggs in different stages of development; the more resistant eggs were 50 hr old and in a tail-free embryo stage while the less resistant eggs were only 17 hr old and in the early gastrula stage of development.

Koo et al. (in prep.) found that American shad eggs acclimated to a base temperature of 24°C could not tolerate exposure to even the lowest ΔT, 10°C, they tested. Hatching success of eggs acclimated to 17°C was not affected by exposure to any of their time-excess temperature histories. Hatching success of eggs acclimated to 18°C was significantly reduced following exposure to ΔT's of 13.0 and 14.5°C. At a ΔT of 11.5°C, hatching success was reduced for an exposure time of 30 min to this maximum ΔT but not following an 8 min exposure. Koo et al. (in prep.) attributed the difference in thermal response of the eggs acclimated to 17 and 18°C, not to the difference of one degree in base temperature, but rather to differences in the stage of development of the eggs at the time of testing. Eggs acclimated to 17°C were over 40 hr old and in the tail-free embryo stage, while the eggs acclimated to 18°C were only 24 hr old and in a late gastrula stage of development.

According to Koo et al. (in prep.) exposure of striped bass eggs acclimated to temperatures as high as 23.5°C did not significantly reduce hatching with any of their time-excess temperature histories. At a base temperature of 24°C however, eggs in approximately the same stage of development did show significantly reduced hatching success following a 20 min exposure to a ΔT of 14.5°C, but not following a 7 min exposure to a ΔT of 13.0°C.

Lauer et al. (1974) tested tolerances of large numbers of striped bass eggs to a combination of temperature elevations for different periods of time and found that "safe temperature exposures" (defined by Lauer et al. as those combinations of ΔT and exposure time that caused no apparent increase in mortality) were less for eggs in early embryonic stages than in later developmental stages. The "safe temperature exposure" for 4 hr old eggs was a ΔT of 6.1°C for up to 60 min, whereas for 36 hr eggs it was a ΔT of 12.0°C for up to 60 min. These differences in thermal sensitivity were also apparent for longer exposure times.

Frank (1973) studied the effects of exposure of carp (*Cyprinus carpio*) eggs acclimated to 25°C and in different stages of development to 10 min periods of acute thermal shock to temperatures of 35, 37.5, 40, 42.5 and 45°C. Eggs exposed to the two highest temperatures did not develop. At the lower temperature survival depended upon the stage of development. Carp eggs were most sensitive to thermal shock during the first six hours of development--cleavage and blastula formation. Exposure of eggs in any stage of development to a temperature of 40°C for 10 min was lethal. During the first two hours of development, exposure to 35°C produced a relatively high frequency of morphologically abnormal larvae. A second period of high sensitivity occurred at 12 to 17 hr after fertilization. This period coincided with blastopore closure and the beginning of organogenesis. The eggs had attained maximum thermal resistance about 33 hr after fertilization.

According to Frank (1973): "For a 10 minute exposure, the thermal limit for newly fertilized eggs was approximately 35°C and for embryos in late stages of development, between 40 and 42.5^{o}C."

The response of different developmental stages of eggs of killifish (*Fundulus heteroclitus*) to a 5 min exposure to 40°C, from a base temperature of 20 to 22°C was studied by Hopkins and Dean (1975). The thermal shock had different effects during the course of development; the earliest stages being some of the most sensitive. Different mechanisms appeared to be involved. In newly fertilized eggs, disruption of cytoplasmic flow caused high mortalities and numerous deformities. Blockage of mitosis at or near cleavage had similar results. Another sensitive period was at about 40 hr (late gastrula and closing of the blastopore), possibly due to disruption of induction, again producing deformities and a high mortality. Bergan (1960) observed blockage of mitosis by temperatures of 39 to 40°C in eggs of the blue gourami (*Trichogaster trichopterus*). The blocking action was reversible upon termination of the shock at these temperatures. At 43°C mitosis was blocked irreversibly.

Unfortunately, none of the experiments described in this section produced data in a form appropriate for the direct construction of thermal resistance curves as shown in Fig. 3.

b. Field studies. We have found no published site studies that are of any significant value in assessing the thermal effects of entrainment on fish egg survival and development.

2. Larvae

a. Laboratory studies. The only laboratory studies the authors are aware of in which larval fishes have been subjected to time-temperature histories characteristic of those experienced during entrainment at a steam electric station are those of Hoss et al. (1973, 1974), Austin et al. (1975), Hoss (in prep.), Schubel et al. (1976), Lauer et al. (1974), Valenti (1974), Coutant and Kedl (1974), Koo et al. (in prep.), Dean (in prep.).

Hoss et al. (1973) exposed larvae of Atlantic menhaden (*Brevoortia tyrannus*), spot (*Leiostomus xanthurus*), pinfish (*Lagodon rhomboides*) and three species of flounder (*Paralichthys spp.*) to excess temperatures of 12, 15, and 18^{0}C for 10, 20, 30 and 40 min periods. Larvae were acclimated to 5, 10, 15 and 20^{0}C, shocked to a higher temperature, held at that temperature for a pre-determined time, then returned immediately to the acclimation temperature.

Hoss et al. (1973) reported that, except for Atlantic menhaden, survival was not significantly affected by exposure to a ΔT of 12^{0}C with any of the acclimation temperatures and exposure times tested. Atlantic menhaden acclimated to 10^{0}C could generally survive an excess temperature of 12^{0}C, but survival of fish acclimated to higher temperatures was significantly reduced following exposure to this, and higher, ΔT's. Survival after a 15^{0}C shock was, for all three species, dependent on acclimation temperature and the duration of exposure; percent survival increased with decreasing acclimation temperature and decreasing exposure period.

Hoss et al. (1974) carefully compared predictions of mortality based on two methods of determining thermal resistance--time-to-death at a constant high temperature and percent mortality after each of a series of "square-wave" time-temperature exposures. They showed greater accuracy and utility when the second method was used. Exposure of all three species to a ΔT of 18^{0}C resulted in significantly increased mortalities for all acclimation temperatures and exposure times, but both of these factors clearly had an effect on survival. For example, over 50% of the spot larvae acclimated to 5^{0}C could survive an 18^{0}C shock at all exposure periods tested. While spot "acclimated to 10°C showed greatly reduced survival to the same shock when exposure was greater than 10 minutes . . ." and "spot larvae acclimated to 15°C could not survive an 18°C shock for any of the exposure periods tested." Hoss et al. (1974) also measured Critical Thermal Maxima, and changes in oxygen consumption of menhaden, spot and pinfish with increasing temperatures.

Hoss and his co-workers (L. C. Coston, personal communication, Jan. 1977) are continuing work with spot (*Leiostomus xanthurus*), pinfish (*Lagodon rhomboides*), and black sea bass (*Centropristis striata*) eggs and hope to conduct thermal shock experiments with larvae of these same species.

Austin et al. (1975) exposed Atlantic silverside (*Menidia menidia*) larvae, reared in the laboratory and acclimated to temperatures of 17, 20, 25 and 30°C, to thermal shock of 8°C and 14°C for 13 min and cooled back to 2°C above the rearing temperature within 15 sec. For base temperatures of 17 and 20°C, and a ΔT of 8°C, no mortalities were observed over the 6 hr period of observation. At a base temperature of 25°C and a ΔT of 8°C, the mortality at the end of 6 hr was 19%. At a base temperature of 30°C and a ΔT of 8°C the mortality at the end of 6 hr was 11%. At a ΔT of 14°C, the mortalities were 3% (base temperature 17°C), 0% (base temperature 20°C), 100% (base temperature 25°C), and 100% (base temperature 30°C). Larvae acclimated to 30°C and exposed to a ΔT of 14°C showed total mortality "immediately" after exposure, Table 4.

Young larvae of the winter flounder (*Pseudopleuronectes americanus*) held at base temperatures of 0, 3, 6, 9, and 12°C were exposed for 13 min to ΔT's of 8, 10, 12 and 14°C and the larvae were then observed for mortalities for up to 96 hr (Valenti, 1974). Only the larvae at 3°C exposed to a ΔT of 14°C yielded mortalities significantly different from controls.

Larvae of another winter spawning fish, the tomcod

TABLE 4 Mortality (%) of Atlantic Silverside (Menidia menidia) *Larvae After Various Thermal Shock Tests (from Austin et al., 1975)*

Acclimation Temp (°C)	No. of Organisms	$\Delta T = +8^{\circ}C$				$\Delta T = +14^{\circ}C$			
		Mortality (%) at Various Times Following Exposure to ΔT							
		t_0^1	$t_{13\ min}^2$	t_b^3	$t_{6\ hr}^4$	t_0^1	$t_{13\ min}^2$	t_b^3	$t_{6\ hr}^4$
17	36	0	0	0	0	3	3	3	3
20	36	0	0	0	0	0	0	0	0
25	36	0	3	3	19	83	94	94	100
30	38	11	11	11	11	100			

t_0^1 ≡ *immediately after initial exposure to* ΔT

$t_{13\ min}^2$ ≡ *13 min after initial exposure to* ΔT*, and the end of exposure to excess temperature*

t_b^3 ≡ *immediately after return to base temperature (acclimation* $+2^{\circ}C$

$t_{6\ hr}^4$ ≡ *6 hr after exposure to* ΔT

(*Microgadus tomcod*) were studied by Lauer et al. (1974). "Safe temperature elevations" (defined by Lauer et al. as those combinations of ΔT and exposure time that caused no apparent increase in mortality) varied with the age of the larvae: 26 hr larvae could withstand exposure to a ΔT of 8.9°C above a base temperature of 1.1°C for 30 min; 44 hr larvae a ΔT of 14.4°C for 30 min; and 400 hr larvae a ΔT of 20°C for 30 min. Using this same criterion of "safe temperature elevation," Lauer et al. examined the effects of 60 min periods of thermal shock on striped bass (*Morone saxatilis*) eggs and larvae. Newly hatched larvae tolerated a ΔT of 3.3°C above a base temperature of 19.7°C and appeared more sensitive than late stage eggs which withstood a ΔT of 14.2°C for 60 min. Older larvae were less sensitive than younger larvae. For one day old larvae the "safe temperature elevation" for a 60 min exposure was 5.7°C; for 10 day old larvae it was 7.3°C; and for 30 day old larvae 12.0°C. Lauer et al. (1974) used their laboratory data to make predictions of the thermal effects of entrainment at the Hudson River (New York) Indian Point Plant.

Coutant and Kedl (1975) conducted thermal bioassays on 2 week old striped bass larvae. From a base temperature of 22°C they survived a ΔT of 7°C for 30 min, but ΔT's of 9°C and 11°C caused approximately 50% mortality with exposures of 5 to 6 min.

Schubel et al. (1976) subjected, in the laboratory, blueback herring (*Alosa aestivalis*), American shad (*Alosa sapidissima*) and striped bass (*Morone saxatilis*) larvae to time-excess temperature histories typical of those experienced by organisms entrained by power plants with a variety of design and operating criteria. The maximum excess temperature (ΔT) ranged from 7 to 20°C above the base temperature (the average surface water temperature on the spawning ground); the time of exposure to a maximum excess temperature from 4-60 min; and the period of cooling back to the base temperature from 60-300 min. In all experiments, initial exposure to the full ΔT was instantaneous. The time-excess temperature curves were similar in form to those shown in Fig. 8.

Schubel et al. (1976) observed that exposure to an excess temperature of 20°C resulted in virtually total mortality of all three species within 2 min of initial exposure. Striped bass larvae were the most temperature tolerant of the three species and could withstand exposure to excess temperatures of up to 10°C for up to 30 min--the longest exposure period tested--with no significant increase in mortality. The thermal response patterns of the other two species were more complicated. Koo and his colleagues are continuing experiments with larvae of these same three species in an attempt to resolve some of the apparent anomalies.

Coutant and Kedl (1975) passed 2 week old larvae of striped bass (*M. saxatilis*) through an isolated condenser tube (without passage through pumps and waterbox). They found that at temperatures below the lethal threshold, passage through the condenser loop at velocities of up to 5.8 m/sec caused no significant increase in mortality over that of the controls. Samples were held at ambient temperatures for some days after the experiments without observing delayed effects of entrainment. When the temperature in the condenser loop was raised to lethal levels of 31.0 and 31.9°C, and the duration of exposure was 5 or 6 min, increased mortalities were recorded. However, control larvae exposed to the same time-excess temperature stress, but without passage through the condenser tube, suffered the same mortalities--suggesting that the mortalities were due to the temperature exposure alone.

In a subsequent report of the continued study (Kedl and Coutant, 1976) larvae of white bass (*Morone chrysops*) survived condenser tube entrainment. Mortalities only increased when the temperature of the tube was at lethal levels, and the thermal stress alone accounted for that mortality. The results tend to substantiate the earlier study of Kerr (1953) at the Contra Costa (Calif.) plant, that passage through the condenser tubes *alone* may not be a large cause of mortality.

Disorientation and cold shock have been observed in laboratory studies. During the course of their abrupt thermal shock studies with larvae of menhaden (*B. tyrannus*), spot (*L. xanthurus*), and pinfish (*L. rhomboides*) acclimated to 5, 10, 15 and 20°C, Hoss et al. (1974) made some observations on the behavior of the larvae during and after thermal shock. Abrupt temperature increases of 15°C and 18°C usually caused immediate reactions; typically erratic swimming, disorientation, violent jumping, and convulsions. If the larvae survived this initial reaction, relatively normal behavior often followed. For a ΔT of 12°C generally little or no distressed behavior was observed. The fish larvae surviving the initial ΔT, received a second thermal shock when they were returned to their initial acclimation temperature. Hoss et al. (1974) observed that reaction to the second shock, the shock associated with abrupt cooling, was often more violent than the first, and appeared more pronounced at lower acclimation temperatures. In many cases larvae apparently little affected in behavior by the initial increase in temperature, were completely immobilized by the sudden decrease in temperature. They suggested that the second shock, a shock which they observed was equivalent to the sudden temperature fall which could occur at a rapid dilution type of power plant discharge, might either acting alone or in conjunction with the first shock of heating, be a direct cause of mortality. Secondarily, it might at least increase the likelihood of severe debilitation and vulnerability to predation.

The suggestion that the secondary cold shock might be the more damaging aspect of the thermal experience of entrainment deserves more investigation. Borrowing again from our knowledge of larger fish, physiological stress responses have been described as more extreme at low temperatures than at high temperatures (Umminger, 1973; Umminger and Gist, 1973).

b. Field studies. Reviews of fish larval mortalities

during power plant entrainment indicate a wide range of mortalities, often as high as 100%. Edsall and Yocum (1972) cite reports of heavy kills of entrained fish larvae at several power plants. It is almost always impossible to unequivocally separate mortalities due to physical, thermal and chemical effects.

An early thermal assessment was made by Kerr (1953) at the Contra Costa (Calif.) plant in which small, 2 to 6 cm, long chinook salmon (*Oncorhynchus tshawytscha*) and striped bass (*Morone saxatilis*) were passed through the condenser tubes of an operating plant. The fish were introduced directly into the condenser tubes, and therefore did not pass through the screens, pumps, or water box. Kerr reported a ΔT of about $16^{\circ}C$, and a total exposure time of 3.5 to 5 min, but he did not report the base temperatures. Kerr's tests with striped bass indicated 94% survival of entrained fish at the end of the 5 day observation period. Behavioral responses of the thermally shocked fish were not reported in any detail nor were they apparently considered in assessing the effects of entrainment, although Kerr did point out that small striped bass "readily go into a state of shock without apparent reason."

Markowski (1959) pumped marked fish through the Roosecote Station, Cavendish Dock (United Kingdom), to determine the effects of entrainment on young fish. He reported that there was no evidence of damage from temperature or chlorine, but that some of the larger organisms suffered mechanical injury. The exposed organisms were held for several days after exposure to observe any delayed effects, but none occurred. Markowski did not report any time-temperature histories, and his data are largely qualitative. A significant fraction of his observations was made during periods of other than peak summer temperatures.

One of the most comprehensive studies on the effects of entrainment on fish larvae was made by Marcy (1971). He reported heavy mortalities of larval fish at the Connecticut Yankee's Haddam Neck plant on the Connecticut River when temperatures in

the plant's 1.83 km long discharge canal exceeded 30°C. The maximum ΔT at the Haddam Neck plant is 12.5°C, and the total travel time to the end of the canal is 50 to 100 min. The heavy mortalities were attributed to "heat shock and prolonged exposure to elevated temperature," but Marcy also reported that "the majority of dead specimens were mangled." In later studies at the same plant, Marcy (1973) reported that in June and July nonscreenable fish were abundant near the plant's intake and that about "80% of the mortality in the canal was caused by mechanical damage and 20% was attributable to heat shock and prolonged exposure to temperatures elevated above 28°C." When canal temperatures were between 29°C and 33.5°C no entrained fish survived to the end of the canal, and at temperatures greater than 35°C, 100% mortality occurred within the plant itself. These results are in contrast to a preliminary study in which Marcy (1969) reported that larval river herring were able to successfully pass through the condenser when the maximum temperature was 34°C.

3. Juvenile Fish

a. Laboratory experiments. In experiments with "half grown" killifish (*Fundulus heteroclitus*), Loeb and Wastneys (1912) established that for fish acclimated to a lower temperature and abruptly placed in water of a higher temperature, the time of survival depended on the size of ΔT. Killifish acclimated to a base temperature of about 10°C died after 4 hr of exposure at 25°C, after 1 hr at 27°C, 13 min at 31°C, and 2 min at 33°C.

Doudoroff (1942) showed that acclimation of juvenile opaleye (*Girella nigricans*) to different base temperatures affected their resistance to acute thermal shock. Doudoroff found 100% survival of juvenile shorefish acclimated to either 20°C or 28°C and exposed to 33°C for 1 hr, but only 25% survival of those fish acclimated to 12°C and exposed to 31°C for 1 hr. In further experiments with young top smelt (*Atherinops affinis*) acclimated to 20°C, Doudoroff (1945) demonstrated how sensitive this species was to small changes in ΔT above some threshold. For ΔT's of 11.5, 12.0, and 12.5°C there was 100% survival for continual exposures of 30 and 60 min. For a ΔT of 13°C survival was 86% for 30 and 60 min exposures, and for an increase in the ΔT of only 0.5°C to 13.5°C, survival was sharply reduced to 42% after a 30 min exposure and to only 17% after a 60 min exposure. From his results for top smelt (*Atherinops*) and opaleye (*Girella*), Doudoroff was able to construct time-temperature graphs for upper median temperature tolerances.

Studies of lethal time-temperature relationships for fish were considerably expanded by Fry et al. (1946), Brett (1952) and co-workers first for goldfish, and later with juvenile salmon and trout. Brett (1952) made extensive studies of juvenile Pacific salmon about 5 cm in length. Median resistance times shown in Fig. 3 for chinook salmon (*Oncorhynchus tshawytscha*) were also determined for juvenile pink (*O. gorbuscha*), sockeye (*O. nerka*), chum (*O. keta*), and coho (*O. kisutch*) salmon, all about 4-5 cm in length. For all five species, the linear fit of the logarithms of the median times to death for different temperatures was highly significant. And, for all five species the slopes of the lines relating median resistance times to temperature were not significantly different. Chinook and coho salmon were significantly more resistant to high temperatures than sockeye, pink, or chum salmon.

Most of Brett's (1952) data were presented in the conventional form of times for 50% mortality, but for young sockeye and chum he also presented data for 10% and 90% mortalities. Young sockeye acclimated to 5°C and raised abruptly to 24°C suffered 10% mortality after 90 min, 30% mortality in 110 min, 50% mortality in 120 min, 70% mortality in 150 min, and 90% mortality in about 190 min. Chum salmon, also acclimated to 5°C, and exposed to acute thermal shock at 23°C, exhibited the following responses: 10% mortality in 65 min, 30% mortality in 80 min,

50% mortality in 100 min, 70% mortality in 120 min, and 90% mortality in 300 min.

The studies of Brett (1956) demonstrated that the upper thermal tolerance of Pacific salmon was near the lower end of the range of values for freshwater fish in temperate North America. The *ultimate*[2] upper lethal temperatures for juvenile coho salmon (*O. kisutch*) and chinook salmon (*O. tshawytscha*) are 25.0°C and 25.1°C respectively. Edsall and Colby (1970) determined the thermal tolerances of juvenile cisco (*Coregonus artedii*) which had been acclimated to different temperatures. Cisco also belong to the Salmonidae. From a base (acclimation) temperature of 25°C, juvenile cisco were exposed to lethal temperatures of from 26.0 to 30.0°C and the times to death recorded for 10% increases in mortality. In other experiments, the median resistance times to abrupt temperature increases were determined for young cisco salmon which had been acclimated to 2, 5, 10, and 20°C. Such detailed information is invaluable in assessing the effects of exposure to time-excess temperature histories typical of those experienced during entrainment by power plants. Comparison of thermal tolerance data for cisco with those for Pacific salmon showed that juvenile cisco were only slightly more tolerant to high temperature than the most tolerant of the Pacific salmon. But, the ultimate upper lethal temperature for juvenile cisco in the laboratory was 26°C whereas the maximum temperature *adult* cisco have tolerated with sustained exposure in the field was only 20°C, 6°C less.

Edsall et al. (1970) tested the thermal tolerance of another coregonine fish, the bloater (*Coregonus hoyi*), to high temperature. Young fish, approximately 6 cm long, were acclimated to base temperatures of 5, 10, 15, 20 and 25°C and then abruptly

[2]*The ultimate lethal temperature is the highest incipient lethal temperature that can be achieved by acclimation.*

exposed to a range of elevated temperatures. The upper lethal temperatures were similar to those found for cisco by Edsall and Colby (1970).

Otto et al. (1976) studied the thermal response of young alewives (*Alosa pseudoharengus*) 2.8-4.7 cm long which had been acclimated to base temperatures of 10-12°C, 18-20°C, and 24-26°C. The fish were subjected to thermal shocks at temperatures ranging from 25 to 34°C. Survival for different time-temperature exposures and median survival times are summarized in Table 5. Young-of-the-year alewives were more tolerant of high temperature than were mature adults of the same species; incipient upper lethal temperatures of juveniles were 3-6°C higher than for adults.

Two small minnows (*Chrosomus eos* and *C. neogaeus*) commonly found in northern temperate river systems including tributaries to the Great Lakes, were tested for resistance to high temperature by Tyler (1966). The minnows, 4-5 cm long, are forage fish. Minnows collected during the winter were acclimated to 9, 15, 20, and 25°C. Minnows collected in summer were acclimated to 6, 9, 15, 20, and 25°C. The minnows were subjected to a range of abrupt temperature elevations and the mortality recorded as a function of time. From these data, median resistance times were calculated for each acclimation temperature. Both species were equally resistant to high temperature when acclimated to 15°C, or higher, but at acclimation temperatures of 9°C, *C. eos* was less tolerant than *C. neogaeus*. At the same acclimation temperature, *C. eos* collected in summer were more resistant to high temperature than those collected in winter.

Hoss et al. (1971) exposed post-larval striped killifish (*Fundulus majalis*) acclimated to 22°C and to a salinity of either 10 ^{o}/oo or 30 ^{o}/oo to ΔT's of 17 and 18°C for periods of 3, 5, 10, and 30 min. After exposure, the fish were returned abruptly to the acclimation temperature. These investigators found that the lethal temperature was a function of both exposure time and

TABLE 5 Survival (%) of Young-of-the-Year Alewives (Alosa pseudoharengus) *Exposed to an Abrupt Increase in Temperature (from Otto et al., 1976)*

	Acclimation Temperature					
	10 - 12°C		18 - 20°C		24 - 26°C	
Test temperature (°C)	Survival (%)	Median survival time (min)	Survival (%)	Median survival time (min)	Survival (%)	Median survival time (min)
25.0	100					
25.5	100					
26.0	70					
26.5	40	150				
27.0	0	15				
27.5	0	14				
28.0	0	9	100			
28.5						
29.0			90		100	
29.5						
30.0			80		90	
30.5			30	170		
31.0			0	70	90	
31.5			0	30		
32.0			0	10	60	
32.5					0	180
33.0					0	76
33.5						17
34.0					0	14
CTM	28.3°C		32.7°C		34.4°C	

salinity. Survival was significantly greater following exposure to the lower ΔT--17°C. Survival at both salinities was significantly reduced for longer exposure periods, 10 and 30 min. Survival following exposure to a ΔT of 18°C was significantly greater at a salinity of 30 $^{\circ}/oo$ than at 10 $^{\circ}/oo$. Doudoroff (1945) had earlier demonstrated for juvenile Pacific killifish (*Fundulus parvipinnis*) that survival time was reduced in direct relation to the size of the ΔT in acute thermal shock.

In nearly all studies with juvenile fishes, investigators reported disorientation and erratic motor behavior following abrupt exposure to a ΔT above some threshold temperature.

b. Field studies. Physical damage to juvenile fishes entrained in the once-through cooling systems of power plants appears to dominate the effects of the other stresses, although an unequivocal assessment of the effects of the individual stresses is rarely, if ever, possible. We know of no field study that permits a completely satisfactory assessment of the thermal effects associated with pump entrainment on any juvenile fish.

In his study at the Contra Costa (Calif.) plant, Kerr (1953) introduced juvenile, 2-6 cm long, chinook salmon (*Oncorhynchus tshawytscha*) directly into the condenser tubes of the operating plant. The fish did not pass through the pumps or the water box. Kerr reported a ΔT of about 16°C, and a total exposure time of 3.5 to 5 min, but he did not report the base temperatures. In ten tests with a total of 100 salmon, Kerr (1953) found that "these fish withstood the 16 degree temperature rise of the condenser with no fatalities." Kerr concluded that for "yearling king salmon of the size able to pass a 3/8 inch mesh screen . . . passage through the plant constitutes no great hazard and survival will be extremely high."

C. Macroinvertebrates

a. Laboratory studies. Embryos and larvae of the hard clam (*Mercenaria mercenaria*) were subjected to 11 different ΔT's for 8 different exposure periods (Kennedy et al., 1974a) and estimates were made of 10%, 50%, and 90% mortalities for different exposure times. The temperature exposures were uniform and continuous and therefore, did not simulate plant entrainment, but larvae were examined for mortalities after 1, 5, 10, 30 min, and longer, so that the data can be related to short exposures to excess temperatures. Because levels of mortality other than the usual 50% were given, the results are more useful than many studies for prediction of entrainment effects. The study shows clearly that the early cleavage stages are the most sensitive, that thermal tolerance increased in trochophore larvae, and that "straight-hinge" stage larvae of the hard clam were even more tolerant. These authors used the data to make predictions of mortality during entrainment. Kennedy et al. (1974b) conducted a similar study of the embryos and larvae of the coot clam (*Mullina lateralis*). The cleavage stages were, as with hard clams, the most sensitive to high temperatures and the "straight-hinge" larvae were least sensitive. Comparing the two species of clams, the cleavage stages of the hard clam were generally more temperature sensitive than those of the coot clam, the trochophore larvae had similar tolerances, and the "straight-hinge" larvae of the hard clam were more tolerant of high temperatures than the larvae of coot clam.

Larvae of the red abalone (*Haliotus rufescens*) were subjected to time-excess temperature tests simulating passage through the Diablo Canyon (Ca.) plant (Adams and Price, 1974). Trochophores acclimated to 17.2°C and subjected to a ΔT of 10°C did not suffer significant mortality. Veliger larvae 41-43 hr old were exposed to ΔT's of 6.7, 10, and 13.3°C for 1 min, without increased mortality. Veligers 61-63 hr old were exposed to a ΔT of 10°C for 1 min and 10 min, and to 13.3°C for 10 min; only those exposed to a ΔT of 13.3°C had significantly higher mortality than the controls.

The Calvert Cliffs (Md.) nuclear power plant on the

Chesapeake Bay has a very low ΔT of $5.6^{\circ}C$. A number of abundant and important crustaceans were tested at time-temperature exposures simulating passage through the plant's condensers (Burton et al., 1976). They were: amphipod (*Gammarus sp.*), shrimp (*Palaemonetes spp.*), mysid (*Neomysis americana*), juvenile blue crab (*Callinectes sapidus*), and mud crab (*Rhithropanopeus harrisii*). The animals were collected at different seasons to take account of the seasonal changes in water temperature. No mortality was observed for any of the five species at any season following exposure to the low ΔT of $5.6^{\circ}C$ for the appropriate time. No sublethal effects were discovered for any of these crustaceans.

Hair (1971) exposed mysid shrimp (*Neomysis awatschensis*) to a series of abrupt temperature increases from 5 to $14^{\circ}C$, for 2, 4, and 6 min followed by a gradual decrease, over a period of 30 min, to ambient temperature. Mysids acclimated to temperatures from 14 to $22^{\circ}C$ showed little mortality until the maximum temperature (ambient plus ΔT) reached $30.5^{\circ}C$; survival above $30.5^{\circ}C$ decreased with increased exposure. In the same study, adult females were examined for effects of temperature shock on reproduction. Females acclimated to $17^{\circ}C$ and exposed to ΔT's of 8.5 and $11^{\circ}C$, later released viable young over a 5 day period, but when acclimated to $21^{\circ}C$ and exposed to a ΔT of $8.5^{\circ}C$, no young were released. The eastern mysid (*Neomysis americana*) was studied by Lauer et al. (1974), who determined 50% and 95% survival values. Mysids acclimated to $25.6^{\circ}C$ and exposed to a ΔT of $7.5^{\circ}C$ for 30 min had 50% survival; 95% of those exposed to a ΔT of $6^{\circ}C$ for 30 min survived. Following exposure to a ΔT of $8.5^{\circ}C$ for 5 min, 50% survived; for a 5 min exposure to a ΔT of $7^{\circ}C$, 95% survived.

Lauer et al. (1974) also made laboratory studies of the thermal tolerance of amphipods (*Gammarus sp.* and *Monoculodes edwardsi*) using the same conditions described above for the mysid (*Neomysis americana*). Both species of amphipods were more temperature tolerant than mysids, and *Gammarus* was the most tolerant of the three species. *Gammarus* showed marked seasonal variations in tolerance to acute thermal shock corresponding to seasonal changes in river temperature. Amphipods (*Gammarus*) collected at a summer base temperature of 25.6°C and exposed to a ΔT of 11.5°C for 30 min had 95% survival; 50% survived a 30 min exposure to a ΔT of 12.2°C. For shorter, 5 min, exposure periods 95% and 50% survived ΔT's of 12.5 and 13.0°C respectively. In parallel tests with the amphipod, *Monoculodes*, 30 min exposures to ΔT's of 9.5 and 10.5°C resulted in survivals of 95% and 50% respectively. With 5 min exposures to ΔT's of 9 and 11.5°C, 95% and 50% of *Monoculodes* survived.

In other thermal shock studies with *Gammarus*, Krog (1954) found that *G. limnaeus* acclimated to 1-3°C and to 15-22°C could withstand 60 min exposures to 26°C and 30-32°C respectively without mortality. Krog (1954) reported that *G. tigrinus* and *G. daiberi* acclimated to 25°C could tolerate exposures to ΔT's of 8.3°C and 10°C from 5 to 60 min with no mortality. *Gammarus sp.* acclimated to a base temperature of 11.7°C survived a 60 min exposure to a ΔT of up to 16.7°C (Ginn et al., 1974; Ginn et al., 1976; Lauer et al., 1974). Ginn et al. (1976) reported that when mature female *Gammarus* acclimated to 26°C were exposed to a ΔT of 8.3°C for 60 min there was no effect either on mating activity, or on the release of young by ovigerous females. *Gammarus sp.* acclimated to 7.1-11.2°C and exposed for a "prolonged" time to a ΔT of 15.6°C were also reported to reproduce successfully.

Stage 1 larvae of the American lobster (*Homarus americanus*) were briefly studied for their thermal tolerance as part of a biocide study by Capuzzo et al. (in press). They found that thermal stress was "not significant" at 20 and 25°C, but that when larvae acclimated to 20-22°C were exposed to a temperature of 30°C for 30 min and 60 min the mortalities were 20% and 30% respectively.

Sprague (1963) applied the acute thermal shock techniques

developed by Fry (1947) and Brett (1952) to studies of thermal resistance in three freshwater crustaceans. Most of the experiments entailed protracted exposures, but at some higher temperatures mortalities occurred in a few minutes. The short-term mortalities are summarized in Table 6. Combining results

TABLE 6 Time in Minutes for 50% Mortality of Freshwater Crustaceans after Acute Thermal Shock (from Sprague, 1963)

	Species							
	Asellus intermedius				Gammarus fasciatus		Hyallela azteca	
	Base Temperature (°C)							
	10	*20*	*25*	*30*	*10*	*20*	*10*	*20*
Shock Temperature (°C)	*Time in Min for 50% Mortality*							
32					*100*			
33	*160*				*12*			
34	*60*					*100*		
35	*23*	*140*				*40*	*60*	
36		*28*	*80*	*150*		*12*	*25*	*80*
37			*40*	*80*				*30*
38			*12*	*35*				
39				*12*				

for the amphipod *Gammarus fasciatus*, the males appeared less temperature sensitive than the females. Size was shown to be a modifying factor of mortality in the isopod *Asellus intermedius* with a general tendency for the larger animals to have shorter survival times. *Asellus* was also studied at different seasons but showed no seasonal changes when acclimated in the laboratory.

b. Field studies. At the Indian Point Plant (N.Y.) on the Hudson River, Lauer et al. (1974) made field tests for comparison with their laboratory studies of macroinvertebrates. When the condenser flow of the plant was throttled to produce a ΔT of 8.3°C, 54% of the mysid (*Neomysis americana*) captured in the discharge were dead. Animals exposed to the same time-temperature exposure histories in the laboratory--in the absence of any physical stress--had a mortality of 50%. Laboratory experiments with the amphipod (*Gammarus*) indicated that it could tolerate the maximum time-temperature exposure in the plant's cooling water system--36.3°C for 32 min--but field observations showed that under those conditions a small, but statistically significant, mortality occurred.

D. Zooplankton

a. Laboratory studies. Heinle (1969) studied mortality of the copepods (*Acartia tonsa* and *Eurytemora affinis*) acclimated to temperatures from 5°C to 25°C when subjected to acute thermal shock. *Eurytemora* had a slightly higher thermal tolerance than *Acartia*. Only the results for the first 60 min of thermal exposure will be considered, because they are within the range of exposure times associated with discharge canals. The results (taken from Heinle's Figs. 1 and 2) are given in Table 7. The upper limits of thermal tolerance for both copepods were found to be near ambient summer temperatures.

Hudson River zooplankton were tested for thermal tolerance by Lauer et al. (1974). The calanoid copepods (*Acartia tonsa*) and (*Eurytemora affinis*) were the most sensitive of the four organisms reported. For a 15 min acute thermal exposure, from a summer base temperature of 24.6°C, 50% survival was reported for *Acartia*, *Halicyclops*, *Bosmina*, and *Eurytemora* at ΔT's of 11.5°C, 13.5°C, 14.5°C and 11.5°C respectively. For the same four organisms, 95% survival for the 15 min exposure was reported at ΔT's of 9.0°C, 12.0, 10.5°C and 9.5°C respectively. *Bosmina* and *Eurytemora* were also tested with a 30 min thermal exposure from a base temperature of 24.6°C. With these longer exposures, *Bosmina* had 50% and 95% survival at ΔT's of 14.5 and 9.5°C respectively;

TABLE 7 Mortality (%) of Copepods Exposed to Acute Thermal Shock. Exposure Time was 60 Minutes Unless Noted Otherwise (from Heinle, 1969).

	Acartia tonsa				Eurytemora affinis				
	Acclimation temp. (^{o}C)								
	5	10	20	25	5	10	15	20	25
Acute shock $\Delta T(^{o}C)$	Mortality (%)								
5	40	0	0	0	0	0	0	0	0
10	0	0	0	45	0	0	15	0	60
15	0	0	100	100	0	0	5	100 (45 min)	100 (15 min)
20	20	0	100	100	0	0	100	100 (30 min)	
25					0	100 (15 min)	100 (15 min)		
30					100 (15 min)	100 (15 min)			
35					100 (15 min)				

the same percentage survival for *Eurytemora* was observed at ΔT's of 11.5°C and 8.7°C respectively. Zooplankton tested for thermal tolerance in winter, spring and fall had mortalities at lower maximum temperatures than summer zooplankton, but the ΔT's tolerated were larger. This is well illustrated by results for (*Eurytemora affinis*) given in Table 8.

*TABLE 8 Critical Time-Temperature Combinations for Survival of the Copepod (*Eurytemora affinis*) Acclimated to Different Ambient Temperatures (from Lauer et al., 1974)*

Ambient temperature (°C)	*Exposure Time (min)*			
	15	*30*	*15*	*30*
	Critical Temperature (°C) for			
	95% Survival		*50% Survival*	
2.4	*28.0*	*28.0*	*30.0*	*30.0*
6.8 to 9.0	*26.0*	*25.0*	*30.5*	*23.5*
12.2 to 15.6	*31.0*	*28.0*	*33.0*	*31.0*
16.1 to 20.6	*31.2*	*31.0*	*34.0*	*33.0*
21.1 to 23.8	*33.0*	*32.0*	*35.0*	*34.0*
25.8	*35.0*	*33.5*	*36.0*	*35.0*

Two species of the freshwater cladoceran *Daphnia* (*D. pulex* and *D. magna*) were studied by Goss and Bunting (1976) to determine their tolerance to acute thermal shock at both higher and lower temperatures. Organisms acclimated to 10, 15, 20, 25, and 30°C were directly transferred to warmer and cooler waters, to a maximum of 30°C and a minimum of 5°C for *D. pulex*, and to a maximum of 30°C and a minimum of 10°C for *D. magna*. Survival exceeded 90% in all cases. Cladocerans were frequently reported to be initially stunned or disoriented in their behavior at ΔT's of 15°C, or more--either higher or lower--and it was several minutes before apparently normal swimming was resumed.

Bradley (1975) also reported initial stunning and disorientation of the copepod (*Eurytemora affinis*) when shocked to a temperature of 34.5^{0}C. Several minutes were required for recovery; the period increasing with the size of the ΔT.

b. Field studies. Zooplankton, such as copepods, cladocerans, and ostracods, are small enough to pass through power plant cooling systems readily and often with apparently low mortalities. Markowski (1959) observed no effect on zooplankton after passage through the cooling condensers of the Cavendish Dock power station (United Kingdom) with a ΔT of 17.5-22.4^{0}C. Many zooplankton have intricate and delicate appendages covered with fine hairs which are involved in a variety of functions including: filter-feeding, sensing, swimming, and reproduction. Thermal, and particularly mechanical, damage to these delicate appendages may lead to an inability of the animal to function normally. E. J. Carpenter (personal communication) considers that subtle mechanical damage to zooplankton during entrainment may be a serious cause of delayed mortality; mortality which has not been often observed because it may not occur until a number of days after entrainment. The studies of Carpenter et al. (1974) and Gentile et al. (unpublished manuscript) support this hypothesis.

Davies and Jensen (1974) examined zooplankton collected at three power plant sites: Lake Norman (N.C.), James River (Va.), and Indian River estuary (De.). Using "zooplankton motility" as an index of survival, these investigators examined a large number of samples and subjected the data to multiple regression analysis. They described the effects of ambient temperature and ΔT on mortality, but noted that there were other factors which sometimes could not be assessed. It is clear that physical and chemical stresses, as well as thermal stresses, affect motility and that therefore it is difficult, if not impossible, to unequivocally assess thermal effects alone at operating power plants. The Indian River plant had a low ΔT of 6^{0}C, and the discharged organisms showed no significant decrease in motility throughout the year. At the James River plant, with a ΔT of $13^{o}C$, motility decreased as the ambient temperature of intake waters increased to the summer maximum. The thermal tolerance of many organisms was exceeded in summer. The Lake Norman plant used a ΔT of $13^{o}C$, but during the winter this was increased to a ΔT of $20^{o}C$, with a concomitant increase in the time of exposure in the plant from 3 to 5 min because of reduced flow. At this plant little decrease in zooplankton motility was found at summer ambient temperatures because the discharge temperature of $33^{o}C$ did not exceed thermal tolerances for most of the zooplankton. In winter however, because of the greater exposure time to excess temperature, motility of organisms was markedly decreased, and presumably mortality was large.

At the Crane power plant (Md.) on Chesapeake Bay, Davies et al. (1976) incubated zooplankton samples, primarily copepods, which had been collected by pump from both the plant intake and discharge. The plankton samples were held in netted containers under continuous flow conditions at three test temperatures: intake plankton were held either at ambient temperature or at the discharge temperature (ambient plus a ΔT of $5.5^{o}C$), discharge plankton were held at ambient temperature, at the discharge temperature (ambient plus a ΔT of $5.5^{o}C$), or at a higher ΔT of about $14^{o}C$. Subsequently containers were cooled at rates typical of the temperature reduction encountered passing through the discharge canal. The samples were then held for 2 weeks at ambient temperature to observe delayed mortalities. Compared with control samples of intake plankton which had been held at ambient temperatures, no significant differences in survival were found for those organisms which had passed through the plant and had then been subjected to the two ΔT bioassays; nor were differences found with survival of the intake plankton transferred to the ΔT of $5.5^{o}C$. Davies et al. (1976) concluded that entrainment at the Crane plant did not significantly affect survival of copepods and other zooplankton.

Detailed field studies of zooplankton were made by Alden et al. (1976) at the Crystal River steam generating plant on the Gulf coast of Florida which has a ΔT of 5.9°C. Six experimental field treatments were set up with copepods in an attempt to separate the various stress factors causing entrainment mortality. The copepods were collected with plankton nets from the intake and discharge areas of the plant. An intake area sample was taken as a control; another intake sample submerged in the discharge area for thermal effects. A discharge sample was assayed for cumulative mortality in plant entrainment; other discharge and intake samples were allowed to drift through the discharge canal--a trip that took two hours. Finally an intake sample was held for two hours in the intake waters as a control for the delayed thermal exposure effects on the drifting discharge canal samples. Samples from each treatment were examined microscopically. The sex and age-class for the major copepod species were noted, as well as survival. The experiments were carried out throughout a year on a biweekly schedule. An overall trend for the discharge populations was that mortalities were relatively low up to a certain threshold temperature, which differed for each species, but was between 30 and 35°C. Above the threshold temperature mortalities almost always rose quickly. An exception was the large copepod, *Labidocera sp.*, which had high mortalities due to mechanical damage, at all temperatures. Alden et al. (1976) concluded that temperature had an obvious and significant effect on small copepod survival, but that each species exhibited different responses to temperature; responses which were modified by salinity.

Alden et al. (1976) had sufficient data to make detailed analyses of the response patterns of the most abundant copepods in the area: (*Oithona sp.*), (*Acartia tonsa*), (*Paracalanus crassirostris*), and (*Euterpina acutifrons*). They separated the mortalities of each sex and life-stage as functions of salinity and temperature (both for the initial thermal shock in the condensers, and for the extended 2 hr exposure at elevated temperatures in the discharge canal). Although there was some tendency for a gradual increase in mortality with temperature in *Euterpina* and *Paracalanus*, in general all four species showed quite low mortalities when exposed to the initial thermal shock of entrainment, except at the very high temperatures experienced during the summer. During passage through the discharge canal mortalities increased, especially at temperatures above 33^{0}C. Comparing the thermal sensitivity of the four species, the coastal marine copepods *Paracalanus* and *Euterpina* were the most sensitive. Even at lower temperatures they showed some tendency for a gradual increase in mortality with temperature. *Oithona* and *Acartia*, estuarine species adapted for life in fluctuating environments, showed little mortality with increasing temperatures until the thermal threshold was closely approached. The estuarine species appeared pre-adapted to tolerate many stresses imposed by power plants, having a higher fitness in this respect than the open coastal species. Usually the juveniles of a species showed lower mortalities than adults, and adult females were usually less sensitive to entrainment effects than were males.

Alden et al. (1976) estimated that mechanical damage to smaller zooplankton accounted for only minor mortalities at the Crystal River (Fla.) plant. They did not however, hold the organisms long enough to observe delayed mortalities that may have occurred. They concluded that an interaction existed between temperature and mechanical injuries which caused higher mortalities than would occur from temperature alone.

V. USING THE DATA IN A CONCEPTUAL MODEL

It is obvious that a conceptual framework, or model, is needed if one is to use thermal tolerance data effectively to predict the thermal effects of entrainment, and to provide guidance in designing power plant cooling systems to reduce thermal stresses to biologically acceptable levels. Thermal resistance curves such as those shown in Figs. 3-5 provide the basis for this framework. The validity of the thermal resistance curve model is well established. Application and interpretation are straightforward, at least conceptually. The usefulness of this model in any particular situation however, depends upon the availability of thermal tolerance data for the important entrainable organisms in an appropriate form--a form that permits construction of thermal resistance curves. Unfortunately, this is rarely the case. There is no shortage of data; only of appropriate data. Many of the thermal stress studies that have been made and continue to be made for power plant applications have little or no predictive value. All too frequently they provide answers in search of questions.

We have attempted to identify the thermal tolerance data for zooplankton, macroinvertebrates, ichthyoplankton, and juvenile fishes that can be used to construct thermal resistance curves. The criteria for acceptability of data were: (1) that exposure to the full ΔT was applied almost instantaneously so that no thermal adaptation could occur, (2) that mortalities were reported as functions of both temperature and exposure time, and (3) that mortalities were reported for a range of exposure times of from a few minutes up to about two hours. If no significant mortality occurred as in Schubel's (1974) experiments, the results are of relatively little value for constructing thermal resistance curves.

If one applies these criteria, the number of studies that can be used to construct thermal resistance curves is relatively small. The sources of pertinent data which we identified are summarized in Table 9. Initially we aggregated the data into four environmental sets--salmonid riverine, lacustrine, open coastal marine, and estuarine. For each of the first three environments we have prepared a single thermal resistance figure with the combined data for: fish eggs, larvae, and juveniles; zooplankton;

and macrobenthos. In the case of the estuarine organisms (for which we had the largest body of data), the data for ichthyoplankton and juvenile fish were plotted on a separate figure from the zooplankton and macroinvertebrates. The data are for 50% mortality. In each figure the data are plotted separately for maximum temperature and for the ΔT experienced, Figs. 9-13. In each figure we have plotted the data without regard for acclimation temperature.

Although this approach is crude, a number of conclusions can be drawn. The slopes of each of the thermal resistance curves for relatively short exposure times indicate that thermal death is indeed a dose response. The relatively gentle slopes of these portions of the curves suggest however, that maximum temperature of exposure is the primary cause of death. After about 20-30 min, mortality is a function of temperature alone. For power plants with exposure times of more than about 20 min, thermal death can be predicted solely on the basis of temperature of exposure. The limit of median lethal temperatures for exposure times of up to 20-30 min appears to be about 27°C for estuarine and coastal marine fish eggs, larvae and juveniles. For lacustrine organisms the limit may be higher at about 30°C; salmonid juveniles appear the most temperature sensitive, with a limit at about 24°C. The wide range of ΔT for 50% mortality in any data set is probably largely a function of the base temperature (acclimation temperature).

The figures indicate that fish eggs and larvae are usually significantly more sensitive to temperature stress than zooplankton and macroinvertebrates found in the same environment. If thermal criteria are set to protect ichthyoplankton, most zooplankton will be adequately safeguarded against thermally induced mortalities.

TABLE 9 Sources of the Thermal

Investigator and Date of Report[a]
Austin, H. M., A. D. Sosnow, and C. R. Hickey. 1975.
Brett, J. R. 1952.
Coutant, C. C. and R. J. Kedl. 1976.
Doudoroff, P. 1942.
Doudoroff, P. 1945.
Edsall, T. A. and P. J. Colby. 1970.
Edsall, T. A., D. V. Rottiers, and E. H. Brown. 1970.
Frank, M. L. 1974.
Heinle, D. R. 1969.
Hoss, D. E., W. F. Hettler, and L. C. Coston. 1974.
Kennedy, V. S., W. H. Roosenburg, H. H. Zion, and M. Castagna. 1974.
Kennedy, V. S., W. H. Roosenburg, M. Castagna, and J. A. Mihursky. 1974.
Lauer, G. J., W. T. Waller, D. W. Bath, W. Meeks, R. Heffner, T. Ginn, L. Zubarik, P. Bibko, and P. C. Storm. 1974.
Schubel, J. R., T. S. Y. Koo, and C. F. Smith. 1976.

Resistance Data Plotted in Figs. 9-13

Organisms and Stages of Development Tested	
Silversides (Menidia menidia)	*Larvae*
Pacific salmons, chinook (Oncorhynchus tshawytscha), *pink* (O. gorbuscha), *sockeye* (O. nerka), *chum* (O. keta) *coho* (O. kisutch)	*Juveniles*
Striped bass (Morone saxatilis)	*Larvae*
Opaleye (Girella nigricans)	*Juvenile*
Top smelt (Atherinops affinis)	*Juvenile*
Cisco (Coregonus artedii)	*Juvenile*
Bloater (Coregonus hoyi)	*Juvenile*
Carp (Cyprinus carpio)	*Eggs*
Copepods (Acartia tonsa & Eurytemora affinis)	*Adults*
Menhaden (Brevoortia tyrannus), *spot* (Leiostomus xanthurus), *pinfish* (Lagodon rhomboides), *flounder* (Paralichtys sp.)	*Larvae*
Coot clam (Mullina lateralis)	*Eggs & Larvae*
Hard clam (Mercenaria mercenaria)	*Eggs & Larvae*
Copepods (Acartia tonsa; Eurytemora affinis; Halicyclops sp.) *amphipods* (Gammarus sp.; Monoculodes edwardsi), *mysid shrimp* (Neomysis americana), *cladocera* (Bosmina longirostris)	*Adults*
Striped bass (Morone saxatilis) *American shad* (Alosa sapidissima) *blueback herring* (Alosa aestivalis)	*Larvae*

TABLE 9

Investigator and Date of Report[a]
Sprague, J. B. 1963.
Tyler, A. V. 1966.

[a]*Complete citations in bibliography at end of chapter.*

VI. MINIMIZING THE THERMAL EFFECTS OF ENTRAINMENT

To minimize mortalities of entrained organisms resulting from *thermal* stresses, it is clear that power plants with once-through cooling systems should: (1) use the lowest excess temperature (ΔT) that is technologically practicable, (2) minimize the transit time through the plant's cooling system and (3) discharge the cooling water into the receiving waters in such a way as to promote rapid mixing--a multi-port diffuser or a jet discharge. Given these operating conditions at any particular power plant, the mortalities of entrained organisms caused *solely by thermal stress* would be minimized for that plant. This is not to say that the *total number* of entrained organisms killed by *all stresses* would be minimized. IT WOULD NOT. In Chapter 6 we develop a very simple model that can be used as a guide in selecting an appropriate ΔT to minimize the total number of organisms killed by all stresses. The choice is dictated largely by the relative importance of the thermal stresses and the physical stresses.

Schubel (1974), Beck and Miller (1974), and Hoss et al. (1974) have suggested that since the effects upon mortality of the physical stresses experienced during entrainment are usually

(Continued)

Organisms and Stages of Development Tested	
Isopod (Asellus intermedius), *amphipods* (Gammarus fasciatus; Hyallela azteca)	*Adult*
Minnows (Chrosomus eos; Chrosomus neogaeus)	*Juvenile*

greater than the effects of the thermal stresses, a power plant should operate at least at the highest ΔT that is biologically acceptable, and perhaps even higher. The highest ΔT that is biologically acceptable for any organism is a function of the magnitude of the ΔT, the ambient temperature, and the time-temperature exposure history. It may be desirable to adjust seasonally the ΔT at a power plant, using higher ΔT's in winter and lower ΔT's in summer. This possibility should be explored further with respect to the thermal tolerances of the juvenile stages of winter spawning and summer spawning organisms.

VII. RECOMMENDATIONS

A. Biological Research

A significant advancement in our ability to predict the mortalities associated with the thermal effects of entrainment and to formulate power plant design and operating specifications can, in our opinion, be attained far more effectively through appropriate laboratory studies than through field studies. We recommend the following approach.

(1) Development of thermal resistance curves such as those

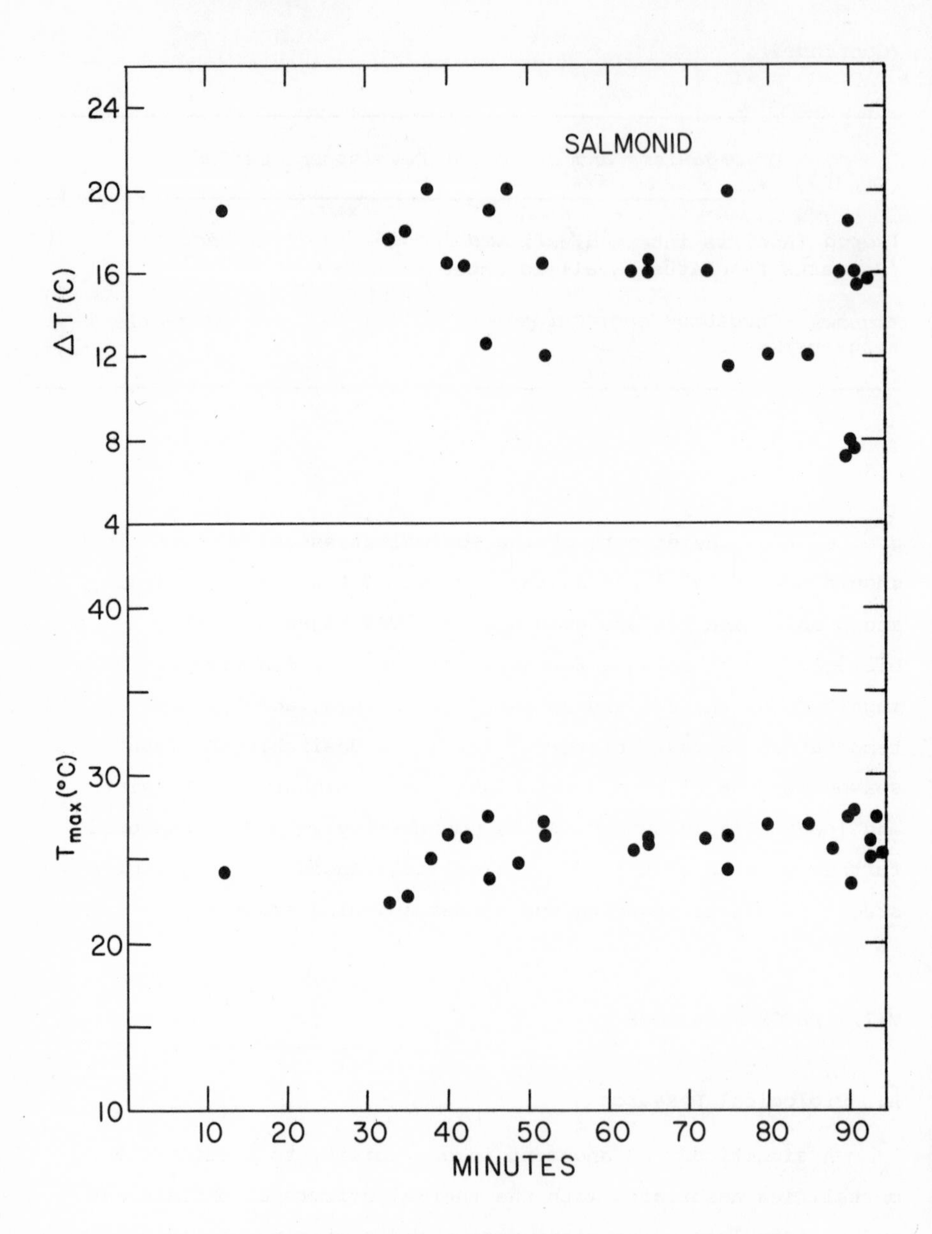

Fig. 9. Community thermal sensitivity. Time to 50% mortality after abrupt exposure to high temperature, T_{max}, (bottom) and to ΔT (top). T_{max} is the actual temperature experienced by the organisms, ΔT is the thermal increment above the base temperature. The data were derived from experiments using different acclimation (base) temperatures.

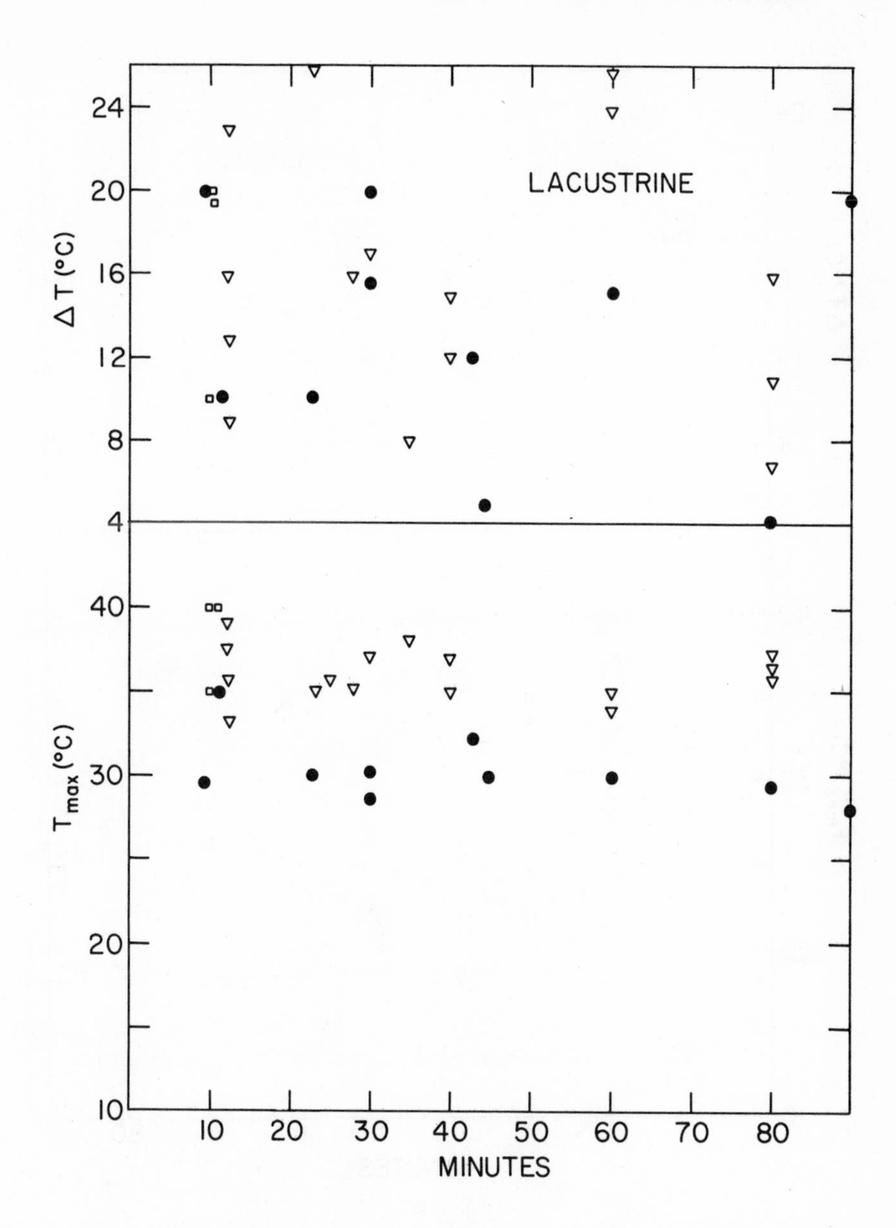

Fig. 10. Community thermal sensitivity. Time to 50% mortality after abrupt exposure to high temperature, T_{max}, (bottom) and to ΔT (top). T_{max} is the actual temperature experienced by the organisms, ΔT is the thermal increment above the base temperature. The data were derived from experiments using different acclimation (base) temperatures.

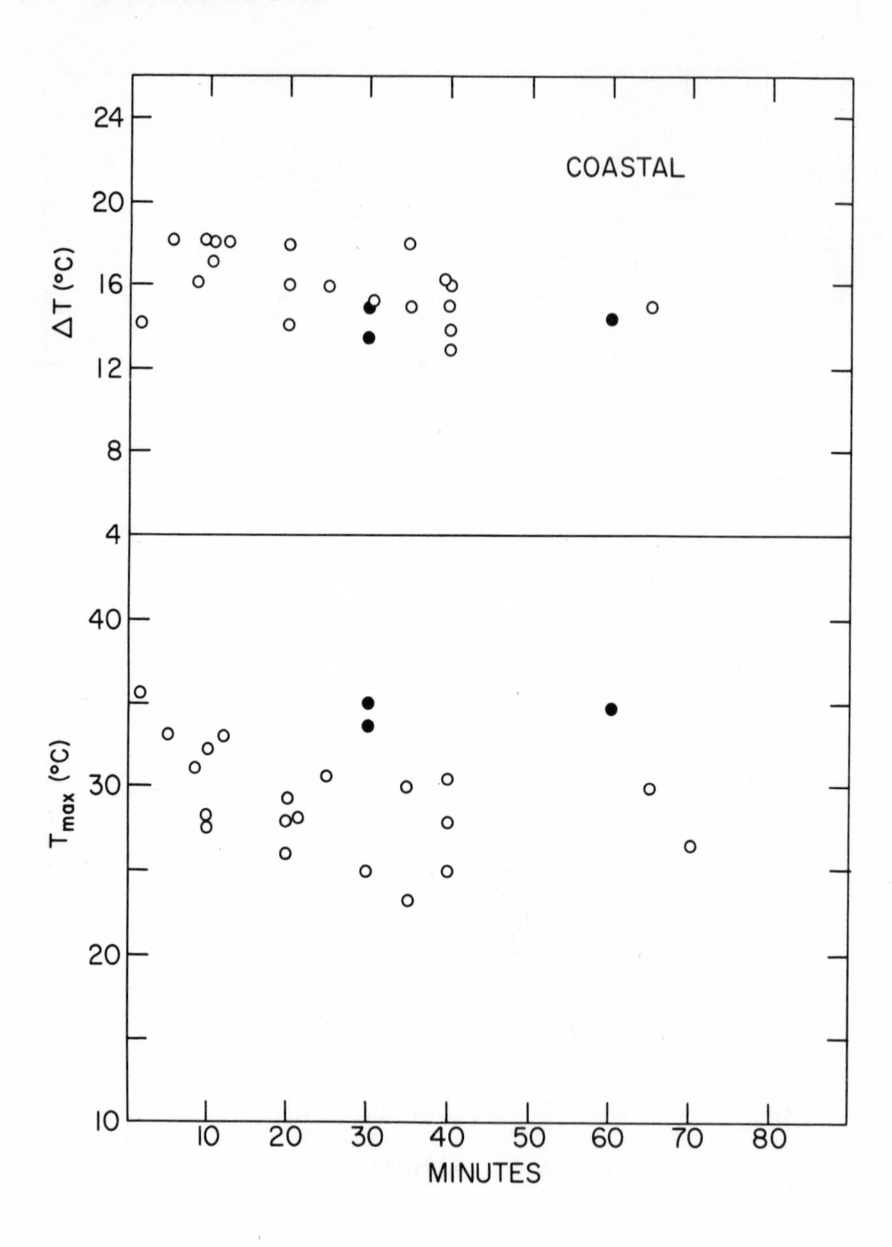

Fig. 11. Community thermal sensitivity. Time to 50% mortality after abrupt exposure to high temperature, T_{max}, (bottom) and to ΔT (top). T_{max} is the actual temperature experienced by the organisms, ΔT is the thermal increment above the base temperature. The data were derived from experiments using different acclimation (base) temperatures.

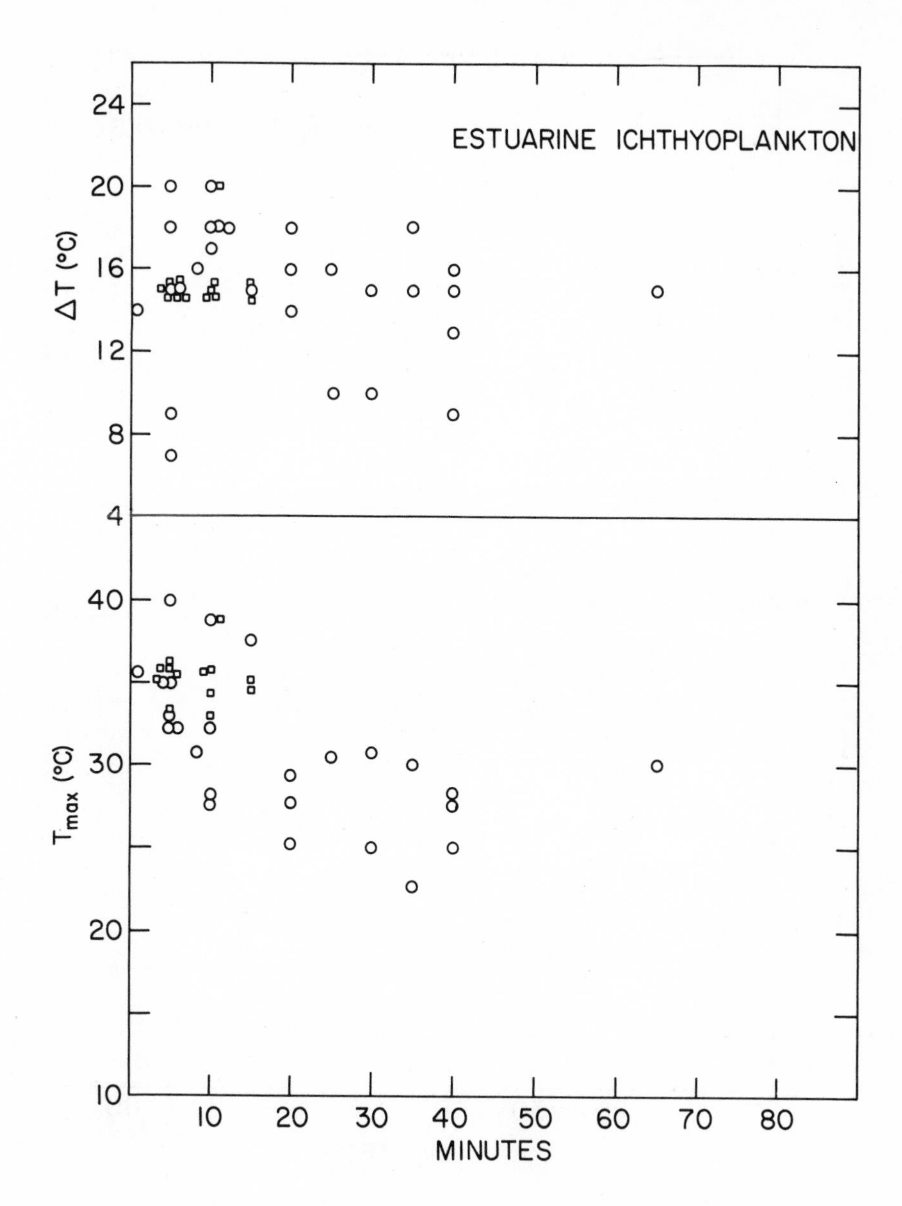

Fig. 12. Community thermal sensitivity. Time to 50% mortality after abrupt exposure to high temperature, T_{max}, (bottom) and to ΔT (top). T_{max} is the actual temperature experienced by the organisms, ΔT is the thermal increment above the base temperature. The data were derived from experiments using different acclimation (base) temperatures.

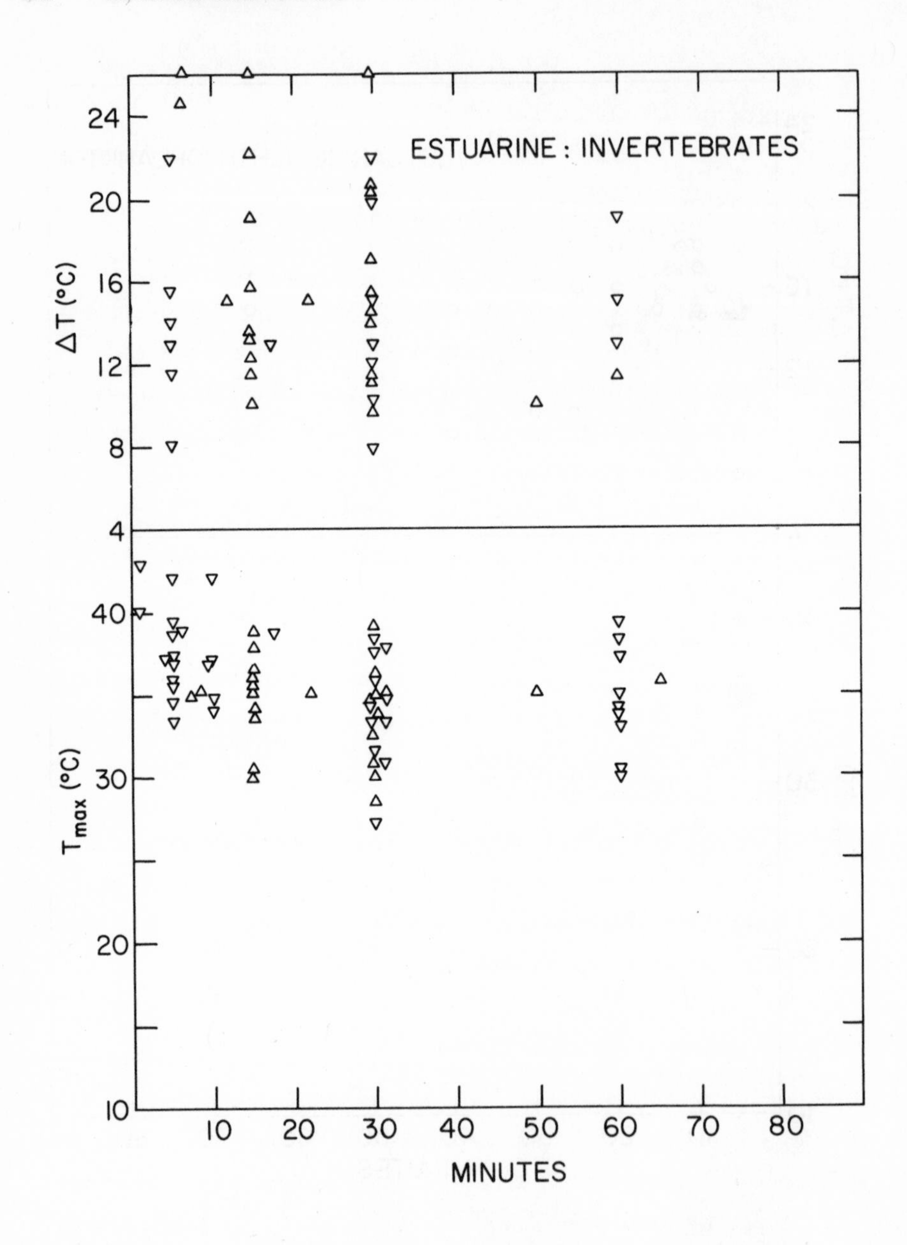

Fig. 13. Community thermal sensitivity. Time to 50% mortality after abrupt exposure to high temperature, T_{max}, (bottom) and to ΔT (top). T_{max} is the actual temperature experienced by the organisms, ΔT is the thermal increment above the base temperature. The data were derived from experiments using different acclimation (base) temperatures.

shown in Figs. 3-5 for a variety of important organisms, particularly ichthyoplankton and juvenile fishes. These curves should be determined for mortalities ranging from 10% to 90% preferrably at intervals of 10%.

(2) Development of representative "community" thermal resistance curves to indicate most sensitive components.

(3) Use of these data with engineering design criteria to develop a variety of time-excess temperature exposure histories that should decrease, to acceptable levels, entrainment mortalities resulting from thermal stresses.

(4) Laboratory verification of the predicted mortalities associated with these time-excess temperature histories, and development of optimal time-excess temperature histories for predictable and acceptable levels of mortality resulting from thermal stress for entrained organisms. Different time-excess temperature histories may be desired at different times of the year for a given power plant site.

(5) Investigation of sublethal effects of acute thermal shock on egg and larval stages, which may disturb development of sensory-nervous systems, or morphogenesis and which may lead to behavior or anatomy inappropriate for survival in subsequent life stages.

B. Power Plant Design and Siting

Thermal damage to entrained organisms can be minimized by a rational approach to siting and design of power plants with once-through cooling systems. Briefly, we recommend the following:

(1) Siting should be based, in large part, on quantitative assessments of relative biological value of potential sites, and on the thermal sensitivity of indigenous organisms, particularly ichthyoplankton and juvenile fishes. Entrainment losses will be less if plants with

once-through cooling systems are located in areas where plankton, particularly meroplankton, is relatively scarce or thermally tolerant, than in more productive areas.

(2) Once-through cooling systems can be designed to minimize entrainment losses if thermal resistance curves for the important species are considered. At an early phase of power plant design, the aquatic organisms to be encountered should be identified, and their survival limitations (discussed above) used for selecting the critical design parameters of ΔT and exposure time in condensers, conduits, and mixing zones. It is clear that the response of aquatic organisms to thermal stress is a dose-response but that the maximum temperature is usually the most important factor in determining thermal death. To minimize the thermal dose experienced during entrainment the transit time through the plant should be minimized and the rate of dilution of the discharge maximized. In most cases, organism survival can be reasonably assured without expensive "overdesign" of the cooling system. Where survival criteria can not be met without costs beyond reason, then the effects of the unavoidable damages to entrained organisms will have to be determined for populations, communities, ecosystems, and man, and decisions made regarding the acceptability of those effects.

(3) Flexible operating modes can often allow protection of organisms during critical times of year while allowing maximum plant efficiency at other times. Design parameters, especially ΔT, can be changed through flexible operation of the number of pumps as the seasonal ambient and discharge temperatures vary.

(4) Power plant design and operating criteria to minimize thermal damage to entrained organisms should be considered in conjunction with probable damages from

physical and chemical stresses to minimize the total number of organisms killed by entrainment (see later chapters).

(5) Once-through cooling is, under certain circumstances, an acceptable mode of power plant cooling, indeed it may be the most desirable mode in some cases.

ACKNOWLEDGMENTS

We thank Alexis Steen and Norman Itzkowitz for their helpful suggestions. We are indebted to Jeri Schoof for tracking down all the goat feathers and for typing the manuscript in final form. The figures were drawn by Carol Cassidy. Contribution 192 of the Marine Sciences Research Center of the State University of New York.

REFERENCES

Adams, J. R. 1969. Ecological investigations around some thermal power stations in California tidal waters. Ches. Sci. 10:145-154.

Adams, J. R. and D. G. Price. 1974. Thermal shock tolerance of larval Red Abalone (*Haliotus rufescens*), from Diablo Canyon, California. Pages 155-175 *in* J. R. Adams and J. F. Hurley, eds. Environmental Investigations at Diablo Canyon 1972-73. Pacific Gas and Electric, Department of Engineering Research, San Ramon, Calif.

Alden, R. W., F. J. S. Maturo, and W. Ingram. 1976. Interactive effects of temperature, salinity, and other factors on coastal copepods. Pages 336-348 *in* G. W. Esch and R. W. McFarlane, eds. Thermal Ecology II. ERDA Symposia Series, Conf. 750425.

Austin, H. M., A. D. Sosnow, and C. R. Hickey. 1975. The effects of temperature on the development and survival of the eggs and larvae of the Atlantic silverside, *Menidia menidia*. Trans. Amer. Fish. Soc. 104:762-765.

Beck, A. D. and D. C. Miller. 1974. Analysis of inner plant passage of estuarine biota. Pages 199-226 *in* Proceedings Amer. Soc. Civil Eng. Power Div. Spec. Conf., Boulder, Colo.

Beltz, J. R., J. E. Johnson, D. L. Cohen, and F. B. Pratt. 1974. An annotated bibliography of the effects of temperature on fish with special reference to the freshwater and anadromous fish species of New England. Massachusetts Agricultural Experiment Station Research Bulletin No. 605, Amherst.

Bergan, P. 1960. On the blocking of mitosis by heat shock applied at different stages in the cleavage divisions of *Trichogaster trichopterus* var. *sumatranus* (Teleostei: Anabantidae). Nytt. Mag. Zool. 9:37-121.

Bradley, B. P. 1975. The anomalous influence of salinity on temperature tolerances of summer and winter populations of the copepod *Eurytemora affinis*. Biol. Bull. 148:26-34.

Brett, J. R. 1952. Temperature tolerance in young Pacific salmon, genus *Oncorhynchus*. J. Fish. Res. Bd. Can. 11:265-323.

Brett, J. R. 1956. Some principles in the thermal requirements of fishes. Q. Rev. Biol. 31:75-87.

Brett, J. R. 1960. Thermal requirements of fish--three decades of study, 1940-1970. Pages 110-117 *in* C. M. Tarzwell, ed. Biological Problems of Water Pollution. Transaction of the 1959 Seminar, U.S. Department H.E.W., R. A. Taft San. Eng. Center, Cincinnati, Ohio. (Tech. Rep. W 60-3.)

Brett, J. R. 1970. Temperature, fishes, functional responses. Pages 515-560 *in* O. Kinne, ed. Marine Ecology, Vol. 1, Part 1.

Burton, D. T., D. B. Richardson, S. L. Margrey, and P. R. Abell. 1976. Effects of low ΔT power plant temperatures on estuarine invertebrates. J. Water Poll. Cont. Fed. 48:2259-2272.

Cappuzo, J. M., S. A. Lawrence, and J. A. Davidson. In press. Combined toxicity of free chlorine, chloramine and temperature to stage I of the American lobster *Homarus americanus*. Water Res.:in press.

Carpenter, E. J., B. B. Peck, and S. J. Anderson. 1974. Survival of copepods passing through a nuclear power station on northeastern Long Island Sound, U.S.A. Mar. Biol. 24:49-55.

Chavin, W. 1964. Sensitivity of fish to environmental alterations. Proc. 7th Great Lakes Conf., Great Lakes Res. Div. Publn. 11:54-67.

Coutant, C. C. 1968. Thermal pollution--biological effects. A review of the literature of 1967 on wastewater and water pollution control. J. Water Poll. Cont. Fed. 40:1047-1052.

Coutant, C. C. 1969. Thermal pollution--biological effects. A review of the literature of 1968 on wastewater and water pollution control. J. Water Poll. Cont. Fed. 41:1036-1053.

Coutant, C. C. 1970a. Thermal pollution--biological effects. A review of the literature of 1969 on wastewater and water pollution control. J. Water Poll. Cont. Fed. 42:1025-1057.

Coutant, C. C. 1970b. Biological aspects of thermal pollution. I. Entrainment and discharge canal effects. C.R.C. Critical Review in Environmental Control 1:341-381.

Coutant, C. C. 1971a. Thermal pollution--biological effects. A review of the literature of 1970 on wastewater and water pollution control. J. Water Poll. Cont. Fed. 43:1292-1334.

Coutant, C. C. 1971b. Effects on organisms of entrainment in cooling water: steps toward predictability. Nucl. Saf. 12:600-607.

Coutant, C. C. 1973. Effects of thermal shock on the vulnerability of juvenile salmonids to predation. J. Fish. Res. Bd. Can. 31:351-354.

Coutant, C. C. and J. M. Dean. 1972. Relationships between equilibrium loss and death as responses of juvenile chinook salmon and rainbow trout to acute thermal shock. U.S. AEC,

Res. Dev. Rept. BNWL-1520, Battelle-Northwest, Richland, Washington.

Coutant, C. C. and C. P. Goodyear. 1972. Thermal effects. J. Water Poll. Cont. Fed. 44:1250-1294.

Coutant, C. C. and R. J. Kedl. 1975. Survival of larval striped bass exposed to fluid-induced and thermal stresses in a simulated condenser tube, ERDA Rept. ORNL-TM-4695, Oak Ridge National Laboratory. 37 p.

Coutant, C. C. and H. A. Pfuderer. 1973. Thermal effects. J. Water Poll. Cont. Fed. 45:1331-2593.

Coutant, C. C. and H. A. Pfuderer. 1974. Thermal effects. J. Water Poll. Cont. Fed. 46:1476-1540.

Coutant, C. C. and S. S. Talmage. 1975. Thermal effects. J. Water Poll. Cont. Fed. 47:1656-1710.

Coutant, C. C. and S. S. Talmage. 1976. Thermal effects. J. Water Poll. Cont. Fed. 48:1487-1544.

Cowles, R. B. and C. M. Bogert. 1944. A preliminary study of the thermal requirements of desert reptiles. Bull. Amer. Mus. Nat. Hist. 83:265-296.

Davies, R. M. and L. D. Jensen. 1974. Entrainment of zooplankton at three mid-Atlantic power plants. *In* L. D. Jensen, ed. Entrainment and Intake Screening Workshop, Feb. 5-9, 1973, The Johns Hopkins Univ., Baltimore, Md.

Davies, R. M., C. H. Hanson, and L. D. Jensen. 1976. Entrainment of estuarine zooplankton into a mid-Atlantic power plant: delayed effects. Pages 349-357 *in* G. W. Esch and R. W. McFarlane, eds. Thermal Ecology II. ERDA Symposia Series, Conf. 750425.

Doudoroff, P. 1942. The resistance and acclimation of marine fishes to temperature changes. I. Experiments with *Girella nigricans* (Ayres). Biol. Bull. 83:219-244.

Doudoroff, P. 1945. The resistance and acclimatization of marine fishes to temperature changes. II. Experiments with *Fundulus* and *Atherinops*. Biol. Bull. 88:194-206.

Edsall, T. A. and P. J. Colby. 1970. Temperature tolerance of young-of-the-year cisco, *Coregonus artedii*. Trans. Amer. Fish. Soc. 99:526-531.

Edsall, T. A., D. U. Rottiers, and E. H. Brown. 1970. Temperature tolerance of bloater, *Coregonus hoyi*. J. Fish. Res. Bd. Can. 27:2047-2052.

Edsall, T. A., T. G. Yocum. 1972. Review of recent technical information concerning the adverse effects of once-through cooling on Lake Michigan. U.S. Fish and Wildlife Service, Bureau of Sport Fisheries and Wildlife, Great Lakes Fishery Laboratory, Ann Arbor, Mich. 86 p.

Esch, G. W. and R. W. McFarlane, eds. 1976. Thermal Ecology II. ERDA Symposia Series, Conf. 750425. 400 p.

Frank, M. 1973. Relative sensitivity of different embryonic stages of carp to thermal shock. Pages 171-176 *in* J. W. Gibbons and R. R. Sharitz, eds. Thermal Ecology. AEC Symp. Series, Conf. 730505.

Fry, F. E. J. 1947. Effects of the environment on animal activity. Univ. Toronto Stud., Biol. Ser., No. 55. Pub. Ont. Fish. Res. Lab. 68:1-62.

Fry, F. E. J. 1967. Responses of vertebrate poikilotherms to temperature. Pages 375-409 *in* A. H. Rose, ed. Thermobiology. Academic Press, N.Y.

Fry, F. E. J. 1971. The effects of environmental factors on the physiology of fish. *In* W. S. Hoar and D. J. Randall, eds. Fish Physiology, Vol. VI, Environmental Relations and Behavior. Academic Press, N.Y.

Fry, F. E. J., J. S. Hart, and K. F. Walker. 1946. Lethal temperature relations for a sample of young speckled trout (*Salvelinus fontinalis*). Univ. of Toronto Stud., Biol. Ser., No. 54. Pub. Ont. Fish. Res. Lab. 66:5-35.

Gentile, J. H., N. Lackier, and S. Cheer. . The effects of entrainment on microzooplankton. (unpublished manuscript).

Gibbons, J. W., and R. R. Scharitz, eds. 1973. Thermal Ecology. AEC Symp. Series, Conf. 730505, Augusta Ga.

Ginn, T. C., W. T. Waller, and G. J. Lauer. 1974. The effects of power plant condenser cooling water entrainment on the amphipod, *Gammarus* spp. Water Res. 8:937.

Ginn, T. C., W. T. Waller, and G. J. Lauer. 1976. Survival and reproduction of *Gammarus* spp. (Amphipoda) following short-term exposure to elevated temperature. Ches. Sci. 17:8.

Goss, L. B. and D. L. Bunting. 1976. Thermal tolerance of zooplankton. Water Res. 10:387-398.

Hair, J. R. 1971. Upper lethal temperature and thermal shock tolerances of the Oppossum Shrimp, *Neomysis awatschensis*, from the Sacramento-San Joaquin Estuary, California. Calif. Fish Game 57:17-27.

Heinle, D. R. 1969. Temperature and zooplankton. Ches. Sci. 10:186-209.

Hopkins, S. R. and J. M. Dean. 1975. The response of developmental stages of *Fundulus* to acute thermal shock. Pages 301-318 *in* F. J. Vernberg, ed. Physiological Ecology of Estuarine Organisms, Belle Baruch Library in Marine Science No. 3.

Hoss, D. E., L. C. Coston, and W. F. Hettler, Jr. 1971. Effects of increased temperature on postlarval and juvenile estuarine fish. Pages 635-642 *in* Proceedings 25th Annual Conference of the Southeastern Assoc. of Game and Fish Comm., Knoxville, TN.

Hoss, D. E., W. F. Hettler, and L. C. Coston. 1974. Effects of thermal shock on larval estuarine fish--ecological implications with respect to entrainment in power plant cooling systems. Pages 357-371 *in* J. H. S. Blaxter, ed. The Early Life History of Fish, Proceedings of an International Symposium. Springer-Verlag, Berlin, Heidelberg, N.Y.

Hutchinson, V. H. 1961. Critical thermal maxima in salamanders. Physiol. Zool. 34:92-125.

Hutchinson, V. H. 1976. Factors influencing thermal tolerances of individual organisms. Pages 10-26 *in* G. W. Esch and R. W. McFarlane, eds. Thermal Ecology II. ERDA Symposia Series, Conf. 750425.

Kedl, R. J. and C. C. Coutant. 1976. Survival of juvenile fishes receiving thermal and mechanical stresses in a simulated power plant condenser. Pages 394-400 *in* G. W. Esch and R. W. McFarlane, eds. Thermal Ecology II. ERDA Symposia Series, Conf. 750425.

Kennedy, V. S., W. H. Roosenburg, M. Castagna, and J. A. Mihursky. 1974a. *Mercenaria mercenaria* (Mollusca: Bivalvia): Temperature-time relationships for survival of embryos and larvae. U.S. Fishery Bull. 72:1160-1166.

Kennedy, V. S., W. H. Roosenburg, H. H. Zion, and M. Castagna. 1974b. Temperature-time relationships for survival of embryos and larvae of *Mullina lateralis* (Mollusca: Bivalvia). Mar. Biol. 24:137-145.

Kennedy, V. S. and J. A. Mihursky. 1967. Bibliography of the effects of temperature in the aquatic environment. Univ. of Maryland Natural Resources Institute, Solomons, Md. 89 p.

Kerr, J. E. 1953. Studies on fish preservation at the Contra Costa Steam Plant of the Pacific Gas and Electric Company. Fish. Bull. No. 92, Calif. Dept. Fish and Game. 66 p.

Kinne, O. 1970. Temperature. Pages 321-616 *in* O. Kinne, ed. Marine Ecology. Vol. 1. Part 1. Wiley Interscience, N.Y.

Krog, J. 1954. The influence of seasonal environmental changes upon the metabolism, lethal temperature and rate of heart beat of *Gammarus limnaeus* (Smith) taken from an Alaskan lake. Biol. Bull. 107:397-410.

Lauer, J., W. T. Waller, D. W. Bath, W. Meeks, R. Heffner, T. Ginn, L. Zubarik, P. Bibko and P. C. Storm. 1974. Entrainment studies on Hudson River organisms. Pages 37-82 *in* L. D. Jensen, ed. Entrainment and Intake Screening, Proc. Second Entrainment and Intake Screening Workshop, Feb. 5-9, 1973, The Johns Hopkins Univ., Baltimore, Md.

Lee, W. S. 1970. Considerations in translating environmental concern into power plant design and operation. Paper presented at annual meeting of Atomic Industrial Forum, Washington, D.C., June 28-30, 1970.

Loeb, J. and H. Wasteneys. 1912. On the adaption of fish (*Fundulus*) to higher temperatures. Jour. Exp. Zool. 12: 543-557.

Marcy, B. C. 1969. Resident fish population dynamics and early life history studies of the American shad in the Lower Connecticut River. The Connecticut River Investigation, 9th Prog. Rept., 13-32.

Marcy, B. C. 1971. Survival of young fish in the discharge canal of a nuclear power plant. J. Fish. Res. Bd. Can. 28: 1057-1060.

Marcy, B. C. 1973. Vulnerability of young Connecticut River fish entrained at a nuclear power plant. J. Fish. Res. Bd. Can. 30:1195-1203.

Marcy, B. C. 1976. Planktonic fish eggs and larvae of the lower Connecticut River and the effects of the Connecticut Yankee plant including entrainment. Pages 115-140 *in* D. Merriman and L. Thorpe, eds. The Connecticut River Ecological Study: The Impact of a Nuclear Power Plant. Am. Fish. Soc. Mongr. No. 1.

Markowski, S. 1959. The cooling water of power stations, a new factor in the environment of marine and freshwater invertebrates. J. Anim. Ecol. 28:243-258.

Needham, J. 1942. Biochemistry and morphogenesis. Cambridge University Press. 785 p.

Otto, R. G., M. A. Kitchel and J. O. Rice. 1976. Lethal and preferred temperatures of the alewife (*Alosa pseudoharengus*) in Lake Michigan. Trans. Amer. Fish. Soc. 105:96-106.

Pickford, G. E., A. K. Srivastava, A. N. Slicher, and P. K. T. Pang. 1971. The stress response in abundance of circulating leucocytes in the killifish *Fundulus heteroclitus*. J. Exp. Zool. 177:89-96.

Pritchard, D. S. and H. H. Carter. 1972. Design and siting criteria for once-through cooling systems based on a first-order thermal plume model. Tech. Report 75, Reference 72-6 of the Chesapeake Bay Institute, The Johns Hopkins Univ., Baltimore, Md. 51 p.

Raney, E. C., B. W. Menzel and E. C. Weller. 1972. Heated effluents and effects on aquatic life with emphasis on fishes, a bibliography. Ichthyological Associates Bull. No. 9, U.S. Atomic Energy Comm. Office of Info. Services, Tech. Info. Center. (Available from Natl. Tech. Info. Serv. Springfield, Va.)

Reaves, R. S., A. H. Houston, and J. A. Madden. 1968. Environmental temperature and the body fluid system of the freshwater Telostii. Ionic Regulation in Rainbow Trout, *Salmo gairdneri*, following abrupt thermal shock. Comp. Biochem. Physiol. 25:849-860.

Rosenthal, H. and D. F. Alderdice. 1976. Sublethal effects of environmental stressors, natural and pollutional on marine fish eggs and larvae. J. Fish. Res. Bd. Can. 33:2047-2065.

Schubel, J. R. S. 1974. Effects of exposure to time-excess temperature histories typically experienced at power plants on the hatching success of fish eggs. Estuarine and Coastal Mar. Sci. 2:105-116.

Schubel, J. R. and A. H. Auld. 1972a. Thermal effects of a model power plant on the hatching success of American shad, *Alosa sapidissima* eggs. Pages 644-648 *in* Proceedings 26th Annual Conference of the Southeastern Assoc. Game and Fish Comm., Knoxville, TN.

Schubel, J. R. and A. H. Auld. 1972b. Thermal effects of a model power plant on the hatching success of alewife (*Alosa pseudoharengus*) eggs. Special Report 28, Ref. 72-14 of the

Chesapeake Bay Institute, the Johns Hopkins University, Baltimore, Md. 13 p.

Schubel, J. R. and A. H. Auld. 1973. Effects of exposure to time-excess temperature histories typically experienced at power plants on the hatching success of blueback herring (*Alosa aestivalis*) eggs. Special Report 31, Ref. 73-4 of the Chesapeake Bay Institute, the Johns Hopkins University, Baltimore, Md. 17 p.

Schubel, J. R. and A. H. Auld. 1974. Hatching success of blue-black herring and striped bass eggs with various time vs. temperature histories. Pages 164-170 *in* J. W. Gibbons and R. R. Sharitz, eds. Thermal Ecology. AEC Symp. Series Conf. 730505.

Schubel, J. R. and T. S. Y. Koo. 1976. Effects of various time-excess temperature histories on hatching success of blueback herring, American shad, and striped bass eggs. Pages 165-170 *in* G. W. Esch and R. W. McFarlane, eds. Thermal Ecology II. ERDA Symp. Series (Conf. 750425).

Schubel, J. R., T. S. Y. Koo, and C. F. Smith. 1976. Thermal effects of power plant entrainment on survival of fish eggs and larvae: a laboratory assessment. Special Report 52, Ref. 76-4 of the Chesapeake Bay Institute, the Johns Hopkins University, Baltimore, Md. 37 p.

Sprague, J. B. 1963. Resistance of four freshwater crustaceans to lethal high temperatures and low oxygen. J. Fish. Res. Bd. Can. 20:387.

Sylvester, J. R. 1971. Some effects of thermal stress on predator-prey interactions of two salmonids. Ph.D. thesis, Univ. Washington, Seattle, Dissertation Abs. 32, 3085-8.

Sylvester, J. R. 1973. Effect of light on vulnerability of heat stressed sockeye salmon to predation by coho salmon. Trans. Amer. Fish. Soc. 102:139-142.

Tyler, A. V. 1966. Some lethal temperature relations of two minnows of the genus *Chrosomus*. Can. J. Zool. 44:349-361.

Umminger, B. L. and D. H. Gist. 1973. Effects of thermal acclimation on physiological responses to handling stress, cortisol, and aldosterone injections in the goldfish, *Carassius auratus*. Comp. Biochem. Physiol. 44:967-978.

Umminger, B. L. 1973. Death induced by injection stress in cold-acclimated goldfish *Carassius auratus*. Comp. Biochem. Physiol. 45A:883-888.

U.S. Environmental Protection Agency. 1973. Water Quality Criteria, 1972. EPA-RIII-73-033, March 1973. 594 p.

Valenti, R. J. 1974. The effects of temperature and thermal shocks on the development of embryos and larvae of the winter flounder, *Pseudopleuronectes americanus*. Final rept. (Contr. No. SR 73-44) to Long Island Lighting Co. 41 p. (mimeo).

Water Resources Council. 1968. The Nation's Water Resources. U.S. Government Printing Office, Washington, D.C. 470 p. (not numbered consecutively).

Wedemeyer, G. 1969. Stress induced ascorbic acid depletion and cortisol production in two salmonid fishes. Comp. Biochem. Physiol. 29:1247-1251.

Wedemeyer, G. 1973. Some physiological aspects of sublethal heat stress in juvenile steelhead trout, *Salmo gaidneri*, and coho salmon, *Oncorhynchus kisutch*. J. Fish. Res. Bd. Can. 30:831-834.

Yocum, T. G. and T. A. Edsall. 1974. Effect of acclimation temperature and heat shock on vulnerability of fry of Lake whitefish (*Coregonus clupeaformis*) to predation. J. Fish. Res. Bd. Can. 31:1503-1506.

CHAPTER 3. BIOCIDES

RAYMOND P. MORGAN, II
EDWARD J. CARPENTER

TABLE OF CONTENTS

I. INTRODUCTION

To optimize heat transfer in condenser systems, fouling organisms, such as bacteria, fungi, algae, protozoans, tube amphipods, barnacles, hydroids, bryozoans, coelenterates, tunicates, mussels, and clams, must either be prevented from entering or periodically removed from intake structures and condenser tubes. Biocides are commonly used to control fouling in the intake and condenser systems of thermal-electric power plants in aquatic environments. In plants which use once-through cooling, chlorine gas (or calcium and sodium hypochlorite) is one of the safer and more economical biocides. However, some plants do use mechanical devices, ozone, or other chemicals (Becker and Thatcher, 1973).

Two special cases need to be noted. First, many power plants use cooling towers. Becker and Thatcher (1973) and Draley (1972) discuss the multiplicity of chemicals employed in cleaning and maintaining cooling towers. The effect of these chemicals will not be considered in this review. Second, many power plants frequently chlorinate their domestic sewage and discharge it into the thermal effluent. In most cases, the discharge flow is very small in proportion to effluent flow. However, any addition of sewage results in increased ammonia, which upon chlorination causes higher concentrations of chloramines.

The chemistry of chlorine as a biocide is complex and is discussed in several references (Sawyer and McCarthy, 1969;

Morris, 1967; Draley, 1972). Hypochlorous (HOCl) and hydrochloric acids (HCl) form during the addition of chlorine to water:

$$Cl_2 + H_2O \rightleftharpoons HOCL + H^+ + Cl^- \quad K_1 = 3.94 \times 10^{-4}$$

Hypochlorous acid further dissociates to form hydrogen and hypochlorite ions (OCl^-):

$$HOCL \rightleftharpoons H^+ + OCl^{-1} \quad K_2 = 3.2 \times 10^{-8}$$

Initially, temperature and pH control the amount of both OCl^- and HOCl. At pH values lower than 4, some free chlorine is present, but at pH 5, 99.5% of the chlorine is present as HOCl and 0.5% is present as OCl^-. The most abrupt change in the ratio of HOCl to OCl^- occurs between pH 7 and 8: at pH 7, HOCl accounts for 72.5% and OCl^- accounts for 27.5% of the chlorine species in solution; at pH 8.0, HOCl accounts for 21.5% and OCl^- for 78.5% of the chlorine species. At a high pH (9.0), OCl^- accounts for 99.0% of the chlorine species in solution. The reactions for both NaOCl and $CA(OCl)_2$ are quite similar.

When any form of chlorine is added to water, initial reactions also occur with organic matter, dissolved gases, and inorganic salts, especially in seawater. Only after this demand is met is there any chlorine residual. In seawater, chlorine chemistry is particularly complex, partially because of halides released by the action of chlorine. Other halogens, especially bromine, then form amines and other derivatives. Recent work by Macalady et al. (1977) on the chemistry of chlorine in estuarine and marine systems indicates that chlorine reacts with naturally occurring bromide and ammonia, furthering the production of hypobromous acid, hypobromite ion and haloamines. Macalady et al. (1977) also note that a phototransformation causes a conversion to bromate, a persistent ionic species.

In many instances, the chemistry of chlorine in freshwater systems is straightforward. However, in some systems such as those carrying high pollution burdens or draining unique soil systems, chlorine chemistry may be as complex as in the estuarine or marine systems.

In addition to reactions associated with chlorine demand, other reactions take place, especially the reaction of HOCl with ammonia. This reaction results in the following products:

$NH_3 + HOCl \rightleftarrows HN_2Cl + H_2O$ monochloramine, $K_3 = 3.6 \times 10^9$

$NH_2Cl + HOCl \rightleftarrows NHCl_2 + H_2O$ dichloramine, $K_4 = 1.33 \times 10^6$

$NHCl_2 + HOCl \rightleftarrows NCl_3 + H_2O$ trichloramine, (little reaction at normal aqueous pH values)

$2NH_2Cl + HOCl \rightleftarrows N_2 + 3HCl + H_2$ nitrogen

The relative amounts of these products are governed by the initial pH and by the ammonia concentration (Merkens, 1958). A low pH coupled with very high HOCl levels leads to formation of more highly substituted derivatives. Until a total residual of chlorines is present, almost all of the chlorine is present as chloramines (Merkens, 1958). Monochloramine is the predominant species when the pH is between 7 and 8 (Marks, 1972); the other two amines are present in only trace quantities (the nitrogen reaction is rare in natural systems).

In any condition of chlorine chemistry, four components must be defined: free residual, combined residual, total residual, and chlorine demand. Free residual chlorine is the hypochlorous acid (HOCl) or hypochlorite ion (OCl^-) remaining after all chlorine demand is fulfilled. Combined residual chlorine is the chlorine which represents reactions with either ammonia or nitrogenous compounds after chlorine demand is fulfilled. Total residual chlorine is the summation of the combined residual and free residual chlorine. These components determine toxicity of the chlorinated water to organisms. An important component of chlorine chemistry in aqueous solutions is chlorine demand. In most cases, chlorine demand is the difference between added chlorine and total residual chlorine for a given temperature and elapsed time, or more simply, the amount of chlorine needed to oxidize all reducing substances in the test water. Any reducing agent found in water contributes to chlorine demand. For example, hydrogen sulfide reacts with chlorine:

$$H_2S + 4Cl_2 + 4H_2O = H_2SO_4 + 8HCl$$

Other ions of concern in freshwater, estuarine, or marine situations include Fe^{++}, Mn^{++}, and NO_2^-. Organics which have a reducing potential or which are unsaturated add to the chlorine demand.

In unpolluted freshwater (<0.03 ppt salinity), chlorine chemistry is simple and is primarily governed by temperature, pH, and ammonia-nitrogen concentrations. In marine waters, chlorine chemistry becomes complicated because of the bromide ions present (approximately 0.08 mM in concentration). The rapid formation of hypochlorite in seawater induces a series of displacement reactions which produce hypobromite and chloride ions (Farkas and Lewin, 1947). At a pH of 8.0 or lower, this reaction is rapid; excess ammonia in the water causes the formation of hypobromites (Johannesson, 1958). When organic material is present in the water, both chlorinated and brominated organics can be formed. Therefore, in any discussion of estuarine or marine chlorine chemistry, toxicity is related to total residual chloro-bromo reactions. In estuarine or marine waters, the constituents of chlorine demand may be toxic. Actually, until the specific components comprising chlorine demand are known, care must be taken in dismissing the "non-toxicity" of these components in estuarine waters. In seawater, chlorine demand is a complex of reactions with chemical components which may be toxic. Refined studies on the nature of those components of chlorine demand in saline systems are needed before analyses of possible impacts can be understood.

Chlorine and calcium or sodium hypochlorite react with a variety of compounds (reviewed in Podoliak, 1974) to form a variety of derivatives, especially chlorinated organics. These chlorinated organics may cause acute hemolytic anemia (Eaton et al., 1973) and may be mutagenic (Shih and Lederberg, 1976). Recent workers have evaluated the relationship between halogenated hydrocarbons in drinking water and cancer mortality (Page et al., 1976, Dowty et al., 1975). Chlorination processes may be potential sources of polychlorinated biphenyls (Walker, 1976).

In any analysis of biocide effects, both intake and plume entrainment need to be considered. Generally, intake entrainment refers to organisms which are taken into and pass through the plant. Plume entrainment refers to organisms which do not pass through the plant but which come into contact with the effluent through turbulent mixing. The effects of biocides are most strongly felt during intake entrainment, however plume entrainment must also be considered because of the sublethal and avoidance effects which can occur.

II. PHYTOPLANKTON

A. Site Studies

Algae are sensitive to chlorine, and the use of chlorine by engineers in the control of algal growth is well known (Mangun, 1929; Campbell and Runger, 1955). However, when used to control fouling organisms in power plants, chlorine injection can also adversely affect the growth of entrained phytoplankton. For example, in a tributary of the Chesapeake Bay, injection of chlorine into cooling water was directly related to the absence of any measurable phytoplankton photosynthesis (Morgan and Stross, 1969). At the same site, photosynthesis was decreased as much as 91% by the addition of chlorine (Hamilton et al., 1970). These studies point up the potentially serious effects of chlorine on entrained phytoplankton.

Later reports cite the detrimental effects of chlorine at other freshwater and marine power generating sites. In Minnesota, Brook and Baker (1972) observed that the productivity of freshwater phytoplankton passing through the plant was depressed 50 to 90% by injection of 2.7 ppm chlorine. At a plant on Florida's west coast, chlorination decreased primary production by 57%. Without chlorination, but with a ΔT of 4.4 to 5.5°C, the depression was only 13%. The residual chlorine within the plant ranged from 0.1 to 1.0 ppm (Fox and Moyer, 1975).

At a power station on upper Narragansett Bay, Gentile et al. (1976) observed that >1.0 ppm total residual chlorine was responsible for complete mortality of all pumped phytoplankton. Dosing was with NaOCl at an initial concentration of about 10 ppm. Total residuals in the discharge canal ranged from 0.5 to 3.0 ppm.

Similar field measurements of photosynthesis were made at the coastal San Onofre plant in California (Eppley et al., 1976). Chlorine was injected at about 1.0 ppm and concentration of total residual in the outfall was between 0.04 and 0.02 ppm. In samples with measurable chlorine, photosynthesis was depressed 70 to 80%. Without chlorine there was no inhibition of photosynthesis.

In a New England coastal power plant, Carpenter et al. (1972) varied the dosage of chlorine and observed the effect on entrained phytoplankton. Chlorine, applied as a gas, was added at concentrations of 0.1 to 1.2 ppm. At the lowest dosage of 0.1 ppm, which resulted in no detectable chlorine residual at the discharge, productivity was depressed by an average of 79%. Since about 0.2 to 0.5 ppm are needed at continuous application to eliminate marine fouling organisms, the authors concluded that there was no safe dosage of chlorine that could be added and yet leave entrained phytoplankton unharmed.

In a Hudson River study, Lanza et al. (1975) investigated the impact of chlorination on phytoplankton assemblages. Water was collected from near the intake structures and from the discharge plume areas and exposed to simulated plume travel. It is not clear from the description of methods how the chlorination episode was duplicated in relation to the simulated discharge plume entrainment. During the studies, the majority of assemblages did not show measurable differences in biomass or productivity between intake and discharge samples except for one measurement where a decrease occurred in a situation with an applied ΔT and chlorine. A later set of experiments indicated that productivity decreased in assemblages exposed to a ΔT with and without chlorination. However, if the water was collected from the discharge plume and used as the experimental material, then the authors did not measure simulated plume entrainment with chlorine. It would seem reasonable to assume the experimental design is measuring a ΔT effect. By the time the samples were prepared for testing, all of the residual chlorine, except for combined forms, would be gone. To further complicate the experimental design, major shifts in phytoplankton populations occurred between the two experiments.

Brooks and Seegert (1977) investigated the effects of intermittent chlorination on phytoplankton from Lake Michigan. In many cases, chlorination resulted in chlorophyll a reductions and phaeophytin a increases. The effect was generally pronounced at chlorine concentrations above 1 ppm. Carbon-14 uptake rates showed drastic reductions following a 30-min chlorine exposure to 0.5 ppm residual chlorine. At chlorine concentrations below 0.1 ppm, there was an initial decrease in ^{14}C uptake ($\sim$20%) but recovery of the stressed phytoplankton population was evident. Chlorine concentrations less than 0.1 ppm did not significantly stress the phytoplankton populations.

B. Viability of Affected Phytoplankton

It appears that phytoplankton affected by chlorine subsequently die. In studies on the Narragansett Bay power plant, Gentile et al. (1976) noted that chains of the diatoms *Detonula confervacea* and *Skeletonema costatum* fragmented and that chlorophyll a decreased an average of 63% as a result of chlorination during the year of study. At the Morgantown plant on Chesapeake Bay, Gentile et al. (1976) noted almost complete destruction of phytoplankton ATP. At the point of dosage, the free residual chlorine concentrations were only 0.55 and 0.32 ppm; in the discharge canal, they were 0.2 and 0.1 ppm. Exposure times were about 2 min in the plant and 45 min in the canal. Since ATP is rapidly lost in cell death, these results indicate irreversible loss of living biomass. Eppley et al. (1976) also noted that there was no recovery of photosynthetic activity in samples even after residual chlorine had fallen to undetectable levels. Thus, rather than merely inhibiting growth, chlorine, as applied in these power plants, apparently acts irreversibly on exposed phytoplankton.

C. Species-specific Sensitivity

Laboratory studies show that algal species differ in their sensitivity to chlorine. The chlorine concentration needed to reduce growth of 11 species by 50% in a 24 hr exposure ranged from 75 to 330 μg Cl_2/l (Table 1), a factor of 4.4 (Gentile et al., 1976). Kott (1969) states that some freshwater algal species, most notably *Cosmarium*, are resistant to chlorine. Similarly, Hirayama and Hirano (1970) observed that *Skeletonema costatum* was killed after exposure to 1.5-2.3 ppm chlorine for 5 to 10 min whereas *Chlamydomonas* was affected at 20 ppm.

D. Dosage-time Effects

As expected, the time of exposure of a phytoplankter to chlorine is critical in determining whether it will be affected. For example, Gentile et al. (1976) noted that there was no measurable mortality of the marine diatom *Skeletonema costatum* at 0.4 ppm chlorine if exposed for about 0.5 min but there was 25 to 50% inhibition of growth if exposed for about 2 to 10 min (Fig. 1). Similar results were obtained for *Thalassiosira*

TABLE 1. *The Toxicity of Chlorine to Selected Species of Marine Phytoplankton*

Species [a]	*24-hr IC-50* [c] *µg/Cl/liter*	*Species* [b]	*24-hr IC-50 µg/Cl/liter*
Skeletonema costatum	95	Chaetoceros decipiens	140
Rhodomonas baltica	110	Thalassiosira nordenskioldii	195
Dunaliella tertiolecta	110	Thalassiosira rotula	330
Monochrysis lutheri	200	Asterionella japonica	250
Thalassiosira pseudonana	75	Chaetoceros didymus	125
		Detonula confervacea	200

[a] *Temperature 20°C.*

[b] *Temperature 10°C.*

[c] *Values are µg/liter producing a 50% growth rate reduction in a 24 hr exposure period.*

pseudonana (Fig. 2). [From these results it is theoretically possible to dose fouling organisms at effective concentrations and, if exposure times are short, to do little damage to phytoplankton.] However, in practice it is extremely difficult to limit chlorine exposure times to a minute or less. Exposure time also influences the effects of chlorination on phytoplankton photosynthesis (Eppley et al., 1976). In working with natural populations, they observed that a 24 hr exposure produced a lower rate of photosynthesis at a given chlorine concentration than a 4 hr exposure (Fig. 3).

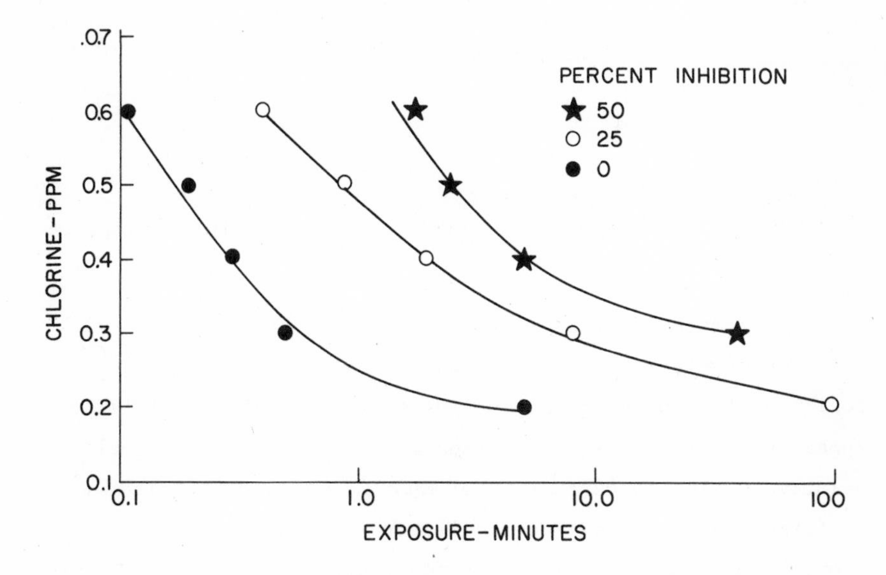

Fig. 1. Response isopleths for the marine diatom Skeletonema costatum *exposed to chlorine. From Gentile et al. (1976).*

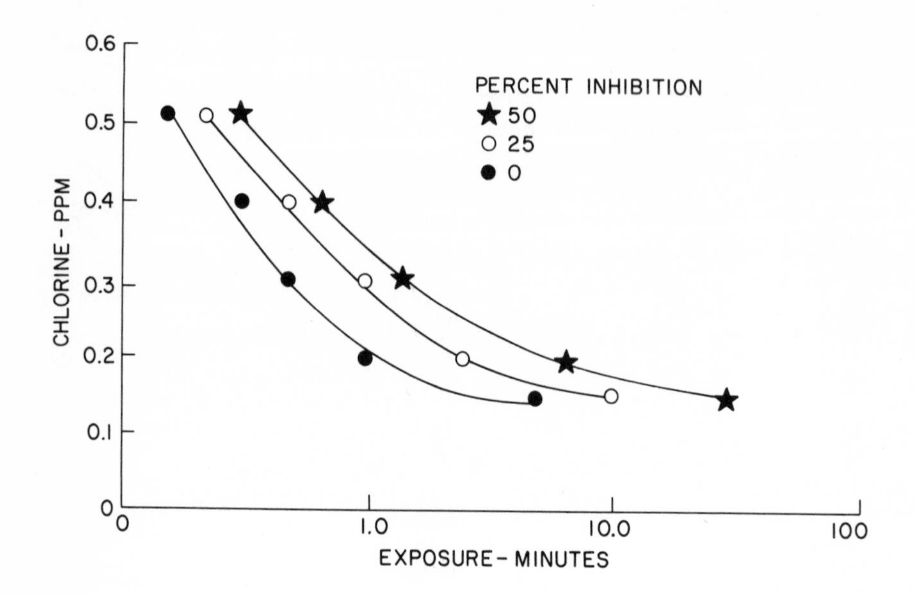

Fig. 2. Response isopleths for the marine diatom Thalassiosira pseudonana *exposed to chlorine. From Gentile et al. (1976).*

E. Complicating Factors

The chemistry of seawater has a major effect on the toxicity of chlorine to phytoplankton. For example, chloramines appear to be more toxic than free chlorine. Thus, if there is a relatively high concentration of ambient NH_4^+, chloramines will be formed and there will be a greater inhibition of algal growth. Eppley et al. (1976) found that a dosage of 0.1 ppm chlorine inhibited phytoplankton production by 50% in 2 to 4 hr incubations (Fig. 3). However, when 30 μM NH_4^+ was added, only 0.06 ppm chlorine was needed to give a 50% decrease in photosynthesis. Roberts (1977) noted a synergetic effect between chlorine and temperature increase on photosynthesis in *Tetraselmis suecica* and natural phytoplankton populations.

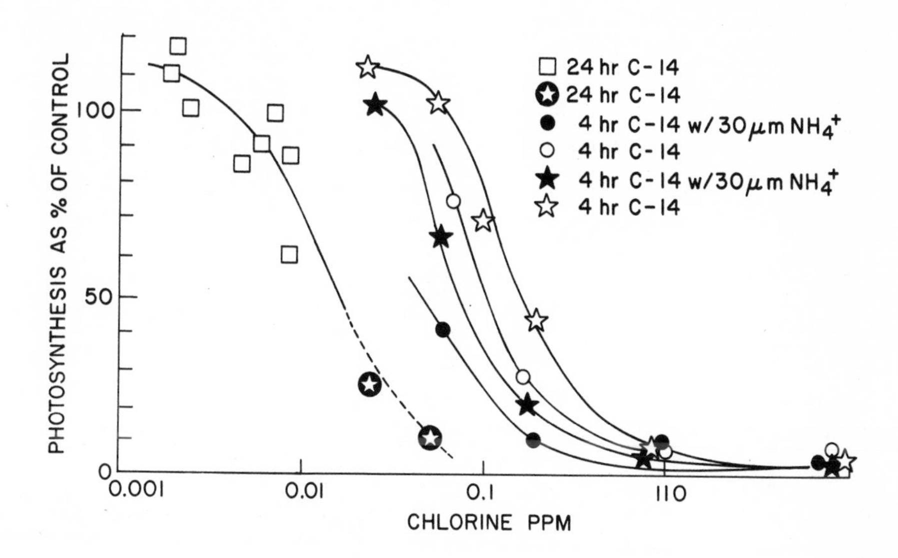

Fig. 3. Inhibition by chlorine of photosynthetic carbon assimilation by natural assemblages of marine phytoplankton. From Eppley et al. (1976).

III. Invertebrates

Chlorine has an adverse effect on many species of entrained invertebrates. For instance, grouping all entrained zooplankton together, Davies and Jensen (1975) observed about 50% zooplankton mortality at 0.25 to 0.75 mg/l chlorine residual at an estuarine plant in Delaware. At 0.50 to 5.00 mg/l, 85 to 100% of the zooplankton were killed.

A. Copepods

The studies concerning chlorine effects on freshwater and marine copepods all indicate marked toxicity. In the Chesapeake Bay area, Heinle (1976) noted high mortality of estuarine copepods because of chlorination. When chlorine was not applied, there was no major effect on copepods. Heinle also found that copepods differed in their sensitivity: *Acartia tonsa* was the most sensitive, followed by *Eurytemora affinis* and *Scottolana canadensis*.

More detailed laboratory studies were carried out on *A. tonsa* by Gentile et al. (1976) since it is a dominant estuarine and marine copepod and appears to be particularly sensitive to chlorine (Fig. 4). It appears from these studies that if exposure time is short (durations less than 5 min) and chlorine concentrations are below about 1.0 ppm, mortality will be low. Heinle and Beaven (1977) found that LC_{50}'s of 0.175 to 0.028 ppm of chlorine produced oxidants (roughly equivalent to chlorine residual, but including all compounds with an oxidizing potential) for adult and copepodid stages of *A. tonsa* (15°C, 10.4-11.8 ppt salinity). However, it should be remembered that in a power plant the effect of chlorine could be magnified by temperature and other stresses. As an example, when *A. tonsa* was exposed to 2.5 mg/l of chlorine and a temperature increase for 5 min at a power plant on the Chesapeake Bay, a mortality of about 90% was observed (McLean, 1973).

The toxicity of free chlorine to the freshwater copepods *Cyclops bicuspidatus* and *Limnocalanus macrurus* was observed by Latimer et al. (1975). The calculated TL_{50} concentrations for 30-min exposures of *Cyclops* at 10, 15, and 20°C were 14.7, 15.6, and 5.8 mg/l, respectively. For *Limnocalanus* at both 5 and 10°C the TL_{50} was 1.54 mg/l. It is apparent for *Cyclops* that a temperature increase from 15 to 20°C markedly increases its sensitivity. The calculated TL_{50} values for 30 min were 0.5 and 0.9 mg/l for *Cyclops* and *Limnocalanus*, respectively. These latter values roughly coincide with concentrations found in most power plants.

At Indian Point, Lanza et al. (1975) reported that *Eurytemora affinis* and *Acartia tonsa* exhibited substantial mortalities when exposed to a simulated plume with a ΔT of

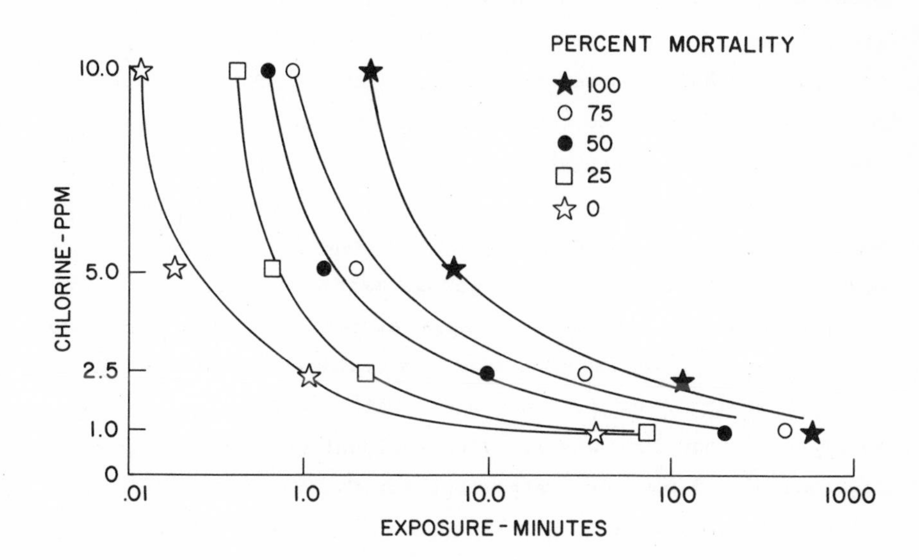

Fig. 4. Response isopleths for the marine calanoid copepod, Acartia tonsa, *exposed to chlorine. From Gentile et al. (1976).*

5.1°C and a condenser residual chlorine of 0.44 mg/l. The chlorine episode duplicated the time-decay pattern of the effluent plume. The combination of temperature and chlorine produced significant mortalities, whereas the simple application of temperature did not. For both *Acartia* and *Eurytemora*, the significant mortalities occurred with the addition of chlorine. Mortalities were monitored 1 hr and 24 hr after exposure.

Apparently, the mode of application of chlorine and other experimental conditions play a major role in the survival of *A. tonsa*. A 96 hr TL_{50} was about 1.0 ppm for a single dose of chlorine at 20°C (Dressel, 1971). These results are not unlike those observed by Gentile et al. (1976). However, if the free residual chlorine dosage is continuously maintained over the observation period, the 48 hr TL_{50} can be very low (below 0.05 ppm) (Roberts et al., 1975). The freshwater copepods *Cyclops bicuspidatus* and *Limnocalanus macrurus* sustained low (<1-5%) mortalities if chlorine concentrations did not exceed about 1.0 ppm for 30 min (Seegert et al., 1977).

Brooks and Seegert (1977) investigated the effects of intermittent application of chlorine to the copepods *Cyclops bicuspidatus thomasi* and *Limnocalanus macrurus*. Bioassays were run for 30 min at a variety of temperatures. For *C. bicuspidatus thomasi* the LC_{50} values were 14.7 ppm residual chlorine at 10°C, 15.6 ppm at 15°C, and 5.76 at 20°C. With *L. macrurus*, the LC_{50} value at 5°C and 10°C was 1.54 ppm residual chlorine. Concentrations of 0.5 ppm were safe (TL_5) for *Cyclops* and 0.9 ppm were safe for *Limnocalanus*.

In a power plant, chlorine is applied as a single dose at the condenser and free residual chlorine disappears within minutes (in clean freshwater, the free residual chlorine may last considerably longer). Thus, the single dose studies of Dressel (1971) and Gentile et al. (1976) probably represent more realistic TL_{50}'s for plant passage than do studies at longer exposure times. Gentile et al. (1976) monitored the free chlorine concentration at the end of the bioassay and reported relatively little loss. In the assays of Roberts et al. (1975), seawater was aerated in test containers and, presumably, this continuously drove off free chlorine, necessitating the constant addition of more. Perhaps the constant introduction of new residual chlorine plus the formation of other chlorine compounds played a role in increasing the toxicity of the chlorine.

B. Amphipods

When exposed to chloramines, the freshwater amphipod *Gammarus* showed marked reduction in the number of young produced as compared with controls. No young were produced in tanks with ≥0.035 mg/l of chloramines during a 15 wk exposure (Arthur and Eaton, 1971). The total number of young produced per adult was markedly less at concentrations of 0.0034 and 0.016 mg/l as compared with controls. The 96-hr TL_{50} value for amphipods was 0.22 mg/l.

Lanza et al. (1975) worked with estuarine gammarids during the Indian Point entrainment study. *Gammarus daiberi, G. tigrinus,* and *Leptocheirus plumlosus* were used in a series of temperature-chlorine avoidance studies. Gammarid survival was generally excellent immediately after plume exposure but survival was reduced when animals were exposed in the discharge canal before being discharged. From the experimental data, it appears that *Gammarus* were able to "sense" chlorinated effluent and to avoid high residual chlorine concentration. Ginn et al. (1974) noted that entrained gammarids exhibited increased initial and latent mortalities during chlorination versus entrainment without chlorination. In normal operation, one-half of the condensors are chlorinated at one time so that the residual chlorine concentration in the plume is significantly reduced. Organisms entrained into the plume did not show the higher mortalities associated with passage through the plant.

Ginn and O'Connor (in press) subjected the amphipod *Gammarus daiberi* to both plume drift and laboratory exposures in order to understand the effect of chlorinated power plant cooling water on mortality. *G. daiberi* survived exposure to the cooling water effluent (ΔT = 7.3-9.3°C, 0.05 ppm total chlorine, 1 hr exposure). In the drifting experiments, no immediate or latent mortalities were observed. Static bioassays for 1 hr indicated an initial TL_{50} value of 1.85 ppm total chlorine. In a series of avoidance experiments, Ginn and O'Connor (in press) found that *G. daiberi* avoided chlorinated discharge water at test temperatures of 26.4-26.6°C and 15.3-15.7°C.

Brooks and Seegert (1977) investigated the effects of intermittent chlorine application on a common lacustrine benthic invertebrate, *Pontoporeia affinis*, and observed LC_{50}'s of 10.6 ppm at 4°C and 3.2 ppm at 9°C (four 30 min exposures over four days). Estimated safe levels (TL_5) were 1.5 ppm residual chlorine at 4°C and 1.4 ppm residual chlorine at 9°C.

C. Grass Shrimp

The shrimp *Palaemonetes pugio* appears to be relatively tolerant of chlorine addition. In tests by Roberts et al. (1975) in a flow-through aquarium system, the TL_{50} was 0.38 ppm after 24 hr and 0.22 ppm after 96 hr. McLean (1973) reported almost no mortality of the grass shrimp exposed to chlorine concentrations between 2.5 and 10.0 ppm for short periods.

D. Barnacles

Nauplii of the barnacle *Elminius modestus* are killed at concentrations of about 0.5 ppm (Waugh, 1964). McLean (1973) found that when barnacle larvae were exposed to 2.5 ppm free chlorine, 75% died in 5 min.

E. Lobster

From tests on Stage I larvae of the lobster *Homarus americanus*, chloramine appears to have a more potent toxic and sublethal effect than equivalent concentrations of free chlorine. Also, a temperature increase enhances chlorine toxicity. Capuzzo et al. (in press, a) exposed the larvae for 30 or 60 min and observed no difference in mortality between exposure times. Significant mortality was noted with exposure to only 0.1 mg free chlorine/l and 0.05 mg chloramine/l. The estimated LC_{50} values were 16.3 and 2.5 mg/l at 25 and 30°C for free chlorine. Chloramine toxicity became more severe as temperature increased, resulting in LC_{50} values of 4.08 mg/l at 30°C. Above 0.50 mg/l mortality increased markedly.

As regards respiration, there was no measurable increase in O_2 consumption during the 30 and 60 min exposures to 0.10 and 1.00 mg/l free chlorine; however, respiration rates increased in the 48 hrs after chlorine exposure and animals exposed to 0.1 and 1.00 mg/l maintained elevated O_2 consumption rates (Fig. 5).

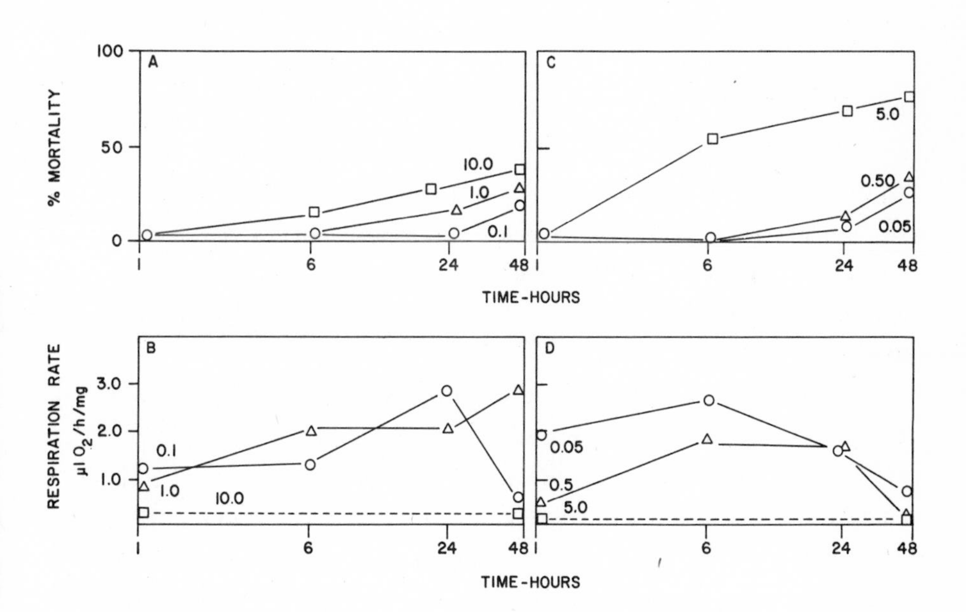

Fig. 5. Percent mortality and respiration rates of stage I lobster larvae detected during the 48-hour period following exposure to applied free chlorine (A & B) and chloramine (C & D). The standard respiration rates of control organisms did not change during this 48-hour period. From Capuzzo et al. (in press, a).

Interestingly, chloramines had a more pronounced effect on respiration than did free chlorine. There was a twofold increase in O_2 uptake when 0.05 mg/l chloramine was applied. However, at higher concentrations, respiration decreased significantly. Changes in respiration were noted within 15 min of exposure to the chloramine. It is clearly apparent that very low concentrations of free chlorine and chloramine have significant lethal and sublethal effects on lobster larvae. The sensitivity is readily noted when we consider that a 60 min exposure of State I larvae to 1.0 mg/l free chlorine or chloramine significantly reduced growth rates during a 19-day observation period (Capuzzo, in press).

F. Molluscs

Bivalve larvae also appear to be sensitive to chlorine but there is some disagreement in the literature. Waugh (1964), for example, states that the larvae of European oyster *Ostrea edulis* are unharmed by as much as 2.5 to 5.0 ppm chlorine for 10 min at 30°C and are capable of growing to the pre-settlement stage.

Roberts et al. (1975), using flow-through systems, found the larvae of the American oyster *Crassostrea virginica* to be very sensitive to chlorine. They obtained an LC_{50} concentration of less than 0.005 mg/l for residual chlorine in a continuous 48 hr exposure. It is possible that the age of the larvae is a major factor affecting sensitivity to chlorine and that this is the reason for the differing results observed by Waugh (1964) and Roberts et al. (1975).

Release of chlorinated effluent from a cooling tower resulted in 50% mortality of snails (*Anculosa* sp.) exposed for 72 hr (30 min each, 4 times/day) at less than 0.04 ppm residual chlorine (Dickson et al., 1974).

G. Other Invertebrates

Chlorinated sewage is a potent spermicide, active in inhibiting fertilization at free chlorine concentrations as low as 0.05 ppm (Muchmore and Epel, 1973). Sperm was obtained from a sea urchin *(Strongylocentrotus purpuratus)*, an echiuroid *(Urechis caupo)*, and an annelid *(Phragmatopoma californica)*. A 5 min exposure to 0.40 ppm residual chlorine reduced fertilization success 100% in the echiuroid and 75% in the annelid. Almost complete failure of fertilization in the sea urchin was noted at an exposure to 0.125 ppm residual chlorine for 5 min.

IV. FISH

A. Field Studies

Biocide effects frequently are difficult to observe in field situations. First, most free chlorine dissipates rapidly in aquatic systems (Dickson et al., 1974) although combined chlorine (the chloramine fraction) requires a substantially longer time to decay (Baker and Cole, 1974). Second, until recently, an instrument capable of measuring low levels of residual chlorine in natural waters did not exist (Marinenko et al., 1976).

Recent review papers discuss the toxicity of chlorine to freshwater fishes (Zillich, 1972), the effect of chlorine on aquatic life (Brungs, 1973, 1976), the use of calcium hypochlorite in fisheries (Podoliak, 1974), the toxicity of power plant chemicals on aquatic life (Becker and Thatcher, 1973), and the effects of sewage effluents on fish (Tsai, 1975). Only a few studies have concentrated on effects of chlorine on freshwater, estuarine, and marine fishes and studies of the effect of chlorine on the early life stages of all fishes are lacking.

In the Patuxent River, Maryland, preliminary information (Mihursky, 1969) on kills of striped bass indicate a possible thermal-biocide episode. However, lack of water quality data and well-documented field observations prevented assessment of the lethal factors. Clark and Brownell (1973) cite an instance when a large number of Atlantic menhaden (*Brevoortia tyrannus*) died at the Cape Cod Canal plant--again, presumably due to chlorine. Truchan (1977) summarized mortalities for chlorinations for five

Michigan plants. Total residual chlorine ranged from 0.01 to 3.05 ppm and the number of fish killed ranged from one to several thousand. Many fish were stressed.

At the Indian Point power complex, the number of dead *Morone* sp. larvae in the discharge increased with chlorination as compared to discharge values without chlorination (Lauer et al., 1974). Residual chlorine levels during the study (measured amperometrically) were approximately 0.11 ppm.

In studies at the Connecticut Yankee plant, Connecticut River, Marcy (1973, 1975, 1976) noted that residual chlorine levels of less than 0.1 ppm contribute the same level of mortality as do mechanical effects, the two being non-separable. Rapid mixing of the injected chlorine and alternate chlorination minimizes chlorine toxicity at the Connecticut Yankee plant.

B. Eggs and Early Larvae

Preliminary analysis of chlorine tolerance data on striped bass *(Morone saxatilis)* indicates that eggs are tolerant to short chlorine exposures, but larvae are susceptible to low chlorine levels (Lauer et al., 1974). Lauer et al. (1974) give little information on methods except to state that the data are preliminary; test conditions, especially salinity and ammonia exposure times, are not given. At Indian Point, striped bass (~63 mm) exposed to a series of temperature-chlorine combinations did not show significant mortality until the "worst possible case" (1 hr exposure, ΔT 6.8°C, 0.16 ppm residual chlorine) of simulated plume entrainment occurred (Lanza et al., 1975). Mortalities ranged from 20 to 50%. However, chlorine in combination with a ΔT produced the same mortalities as a ΔT alone.

The 48 hr TL_{50} for early larvae of plaice *(Pleuronectes platessa)* was 0.032 ppm of free chlorine but there was no effect on eggs at chlorine levels of 0.75 ppm in an eight day exposure (Alderson, 1972). In further studies with plaice and Dover sole *(Solea solea)*, Alderson (1974) found that newly-hatched larvae were more sensitive to chlorine than eggs. Before metamorphosis, sensitivity decreased with increasing age of the larvae. A variety of different stages of plaice and Dover sole have been studied for many durations of exposure by Alderson (1974). Generally, the LC_{50} values, except for eggs of the plaice, were below 0.10 and above 0.025 ppm of chlorine for 48 and 96 hr exposures, respectively.

Middaugh et al. (1977) found incipient LC_{50} values of 0.04 ppm residual chlorine for two-day-old striped bass larvae, 0.07 ppm for 12-day-old larvae and 0.04 ppm for juveniles. Gills and pseudobranchs were damaged when larvae were exposed to 0.21 to 2.4 ppm residual chlorine.

Morgan and Prince (1977) observed LC_{50} values ranging from 0.20 to 0.40 ppm residual chlorine for eggs of five species and 0.20 to 0.32 ppm residual chlorine for larvae of three species. Age-related effects were noted in both egg and larval stages.

Other chlorinated products possibly produced in power plants may have sublethal effects on carp *(Cyprinus carpio)* eggs. Gehrs et al. (1974) and Eyman et al. (1975) note that 4-chlororesorcinol and 5-chlorouracil, identified by Jolley (1975) in sewage water, decreased the hatchability of carp eggs when concentrations of the two compounds were near 0.001 ppm (Gehrs et al., 1974). Additional work by Eyman et al. (1975) indicates that exposure to 0.5 ppm or more of 5-chlorouracil produced abnormal larvae.

C. Juvenile and Adult Fish

Stober and Hanson (1974) note that pink *(Oncorhynchus gorbuscha)* and chinook *(O. tshawytscha)* salmon became less tolerant to increasing temperature and exposure time when exposed to residual chlorine. The LC_{50} values for 2 hr exposures were approximately 0.045 ppm. Chlorine levels for this study were determined with the orthotolidine method, a technique with doubtful accuracy, precision, and sensitivity in the ppb range, especially in saline waters.

TL_{50} values for a variety of estuarine species indicate sensitivity to residual chlorine at a variety of exposures (Roberts et al., 1975). For gobies *(Gobiosoma bosci)* and silversides *(Menidia menidia)*, TL_{50} values of 0.080 to 0.095 ppm are normal at a 24 hr exposure. Pipefish *(Syngnathus fuscus)* have TL_{50} values around 0.28 ppm. Eren and Langer (1973) find that larger Mozambique mouthbrooders *(Tilapia Mossambica)* are more sensitive than the above fish to residual chlorine; free chlorine is more toxic than combined chlorine. Increasing temperatures caused greater sensitivity to chlorine.

Gentile et al. (1976) found that LC_{50} values were approximately 0.10 to 0.20 ppm for yellowtail flounder *(Limanda ferruginea)* for a 24 hr exposure. LC_{50} values for winter flounder *(Pseudopleuronectes americanus)* for 90 and 70 min exposures were 0.100 and 0.130 ppm, respectively. For juvenile menhaden the LC_{50} for a 60 min exposure was 0.22 ppm. Gentile et al. (1976) conclude that for the protection of entrained larvae and for organisms in the receiving water, chlorine levels (residual) should be less than 0.1 ppm.

Capuzzo et al. (in press, b) report on the differential effects chloramine and free chlorine have on juveniles of three fish, killifish *(Fundulus heteroclitus)*, scup *(Stenotomus chrysops)*, and winter flounder. Generally, the responses of the three species to free chlorine were similar, but killifish were more susceptible to chloramines than the other two species. Below lethal levels, behavioral changes such as gill distention and abnormal swimming behavior were common.

Basch and Truchan (1974) report the ILC-50 (intermittent lethal concentration) value for brown trout *(Salmo trutta)* to be approximately 0.56 ppm (96 hr exposure, 21°C). In a series of caged experiments at five Michigan power plants, brown trout and fathead minnows *(Pimephales promelas)* had ILC-50 values of 0.14 to 0.19 ppm intermittent residual chlorine concentrations. Brown trout (48 hr exposure) were subjected to two and four 30 min chlorinations. (Fathead minnow data were incomplete due to significant mortalities among the control fishes.)

Brooks and Seegert (1977), in studying the effects of intermittent chlorination, exposed coho salmon *(Oncorhynchus kisutch)*, rainbow trout *(Salmo gairdneri)*, yellow perch *(Perca flavescens)*, alewife *(Alosa pseudoharengus)*, spottail shiner *(Notropis hudsonius)* and rainbow smelt *(Osmerus mordax)* to a variety of chlorine-temperature exposure regimes. These results are detailed in Tables 2 and 3. In general, at the lower temperatures, LC_{50} values for these fishes were quite large.

Heath (1977) exposed fingerling rainbow trout *(S. gairdneri)*, coho salmon *(O. kisutch)*, golden shiners *(Notemigonus crysoleucas)*, carp *(Cyprinus carpio)* and channel catfish *(Ictalurus punctatus)* to both free chlorine and monochloramine. The method of dosing attempted to duplicate conditions similar to those observed in the field. Rainbow trout, coho salmon and channel catfish were sensitive to both chlorine forms. Free chlorine was more toxic to the tested species than monochloramine (3-14 times greater).

Mattice and Zittel (1976) review the considerations for site-specific evaluation of power plant chlorination and substantially summarize both the existing toxicity data for freshwater and marine organisms and the feasibility of the various techniques for measuring chlorine (Fig. 6). Generally, they conclude that 1) measurement of total residual chlorine adequately defines toxicity in natural waters, 2) acute and chronic tests should be used (flow through systems), 3) freshwater organisms appear to be more sensitive to chronic chlorine exposures, and 4) marine organisms are more sensitive to acute exposures.

Avoidance responses to chlorine are also important in the overall reaction of organisms to biocides. White (1972) and Sprague and Drury (1969) report avoidance behavior for fishes in discharge plumes containing residual chlorine concentrations of

TABLE 2. LC_{50} (in ppm residual chlorine) values (30 min exposure) for Lake Michigan fishes subjected to chlorine (from Brooks and Seegert, 1977)

Species	Temperature				
	10°C	15°C	20°C	25°C	30°C
Yellow perch	8.0	3.9	1.11	0.97	0.70
Rainbow trout	0.99-2.11[a]	0.94[b]	0.43[a],0.60[c]		
Coho[d]	0.56	1.38	0.29		
Alewife	2.15	2.27	1.70	0.96	0.30
Spottail shiners	1.4,2.4[e]	1.00	0.54		
Smelt	1.2				

[a]Size dependent
[b]"Sensitive"
[c]"Resistant"
[d]1976 data
[e]March versus July exposure

TABLE 3. LC_{50} values (in ppm residual chlorine) for Lake Michigan fishes exposed to three 5-min chlorine doses (from Brooks and Seegert, 1977)

Species	Temperature	
	10°C	20°C
Yellow perch	22.6	9.0
Rainbow trout	2.87	0.82[a],1.65[b]
Smelt	3.3	

[a]Sensitive
[b]Resistant

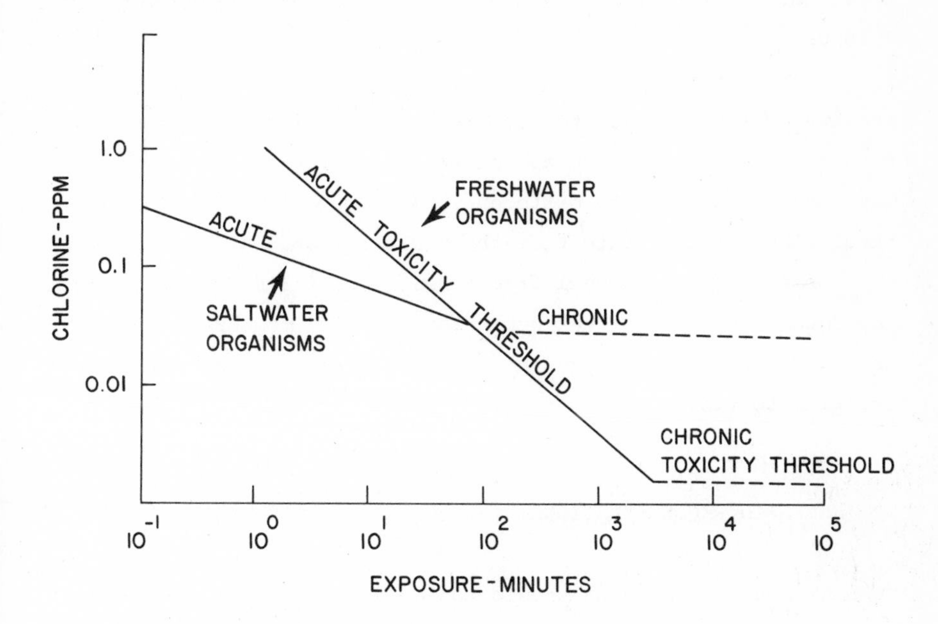

Fig. 6. Acute and chronic toxicity thresholds for marine and freshwater organisms exposed to chlorine.

0.01 to 1.0 ppm (a preference response occurs at 0.1 ppm). Fava and Tsai (1976), while working with sewage, found that blacknose dace *(Rhinichthys atratulus)* responded quite differently to combined versus free chlorine. The free chlorine elicited a lesser degree of fish avoidance and fish exposed to the preference concentrations (0.19 and 0.52 ppm) died in 5 and 2 to 4 hrs, respectively (Tompkins and Tsai, 1976). Meldrim et al. (1974) determined that low levels (0.15-0.04 ppm) of free chlorine caused avoidance in estuarine fishes. Experimental problems did not allow definite conclusions for avoidance reactions. More recently, Middaugh et al. (1977) noted reproducible avoidance reactions by 24-day-old striped bass larvae to total residual chlorine concentrations of 0.79 to 0.82 ppm and 0.29 to 0.32 ppm. No measurable avoidance was observed between 0.16 and 0.18 ppm.

Cherry et al. (1977) evaluated the responses of a variety of freshwater fish species to temperature and chlorine exposures. Considerable variation in avoidance and preference to chlorine was observed along with differences due to acclimation temperature. The authors state that there are few studies on avoidance which emphasize the actual fractional composition of the chlorine residual. It appears that hypochlorous acid, temperature, and pH significantly influence the avoidance response and toxicity of chlorine to fish.

V. RESEARCH RECOMMENDATIONS

Obviously, research should continue on the chemistry of chlorine in freshwater, estuarine, and marine waters. Of primary importance is a better understanding of the synergistic factors, such as pH and temperature, governing the ratio between residual and combined chlorine. It appears that, in marine waters, bromine governs the ionic species accounting for both biocidal activity and residual toxicity. In estuarine waters, all aspects of chlorine chemistry in a variety of estuarine regimes need to be investigated. Concurrent with initial understanding of chlorine chemistry in a variety of waters is the determination of exactly which chlorinated or brominated organics are produced by power plant chlorination. If some of the halogenated products are found to be either carcinogenic, mutagenic, or teratogenic, it may then be necessary from a human health standpoint to discontinue chlorination. Of particular importance to the receiving waters is the impact of chlorination on community processes and structure in both the near-field and far-field. Organisms which are sensitive to chlorination products and the biological changes occurring at a variety of sites must be further investigated. Studies on the bioaccumulation and biomagnification of halogenated organics produced by power plant chlorination should be made in conjunction with near-field studies.

Some research should focus on the applicability of monitoring concentrations of xenobiotics in gall bladders of fishes exposed to power plant effluents. This technique may be very applicable since it appears to have some use with compounds of low lipid solubilities which typically have low bioaccumulation potentials in fatty or live tissue (Stalham et al., 1976). The technique may be of greatest use in systems where high organic loading produces a variety of chlorinated organics.

Finally, research on avoidance and preference needs to be completed for a variety of organisms, especially invertebrates. Work in this area should focus on the levels of biocides which cause both attraction and avoidance.

VI. SUMMARY AND RECOMMENDATIONS

From both field and laboratory studies, it is apparent that chlorine added to aquatic systems seriously affects the growth and survival of entrained organisms. There is species-to-species variation, but the majority of species within a group appear to be affected at relatively low concentrations of residual chlorine (less than 0.5 ppm). In some cases, such as phytoplankton, the action of chlorine may be irreversible. In entrained animals, the effect of chlorine may be manifested by abnormalities in growth or reproduction and by other sublethal expressions. The chemistry of chlorination is simple in uncontaminated freshwater systems whereas estuarine and marine systems are complex because of the inherent chemical composition of these waters. Of primary importance are the chemical characteristics of the ambient water prior to chlorination, especially the levels of ammonia and organic material.

U.S. Environmental Protection Agency (EPA) effluent guidelines for power plants allow levels of chlorine which are inadequate for protection of biota, both during entrainment and in the receiving water. Current guidelines specify that free chlorine residuals shall at no time exceed an average of 0.2 mg/l or a maximum of 0.5 mg/l at the discharge unit and that the discharge time shall not exceed 2 hr per day from a unit.

At present (based on information from Mattice and Zittel, 1976) freshwater releases (on a continual basis) of residual chlorine should not exceed 1.5 ppb for maximum protection (i.e. chronic toxicity threshold) of the organisms in the receiving waters. For marine systems, the chlorine residual should not exceed 20 ppb. For estuarine waters, due primarily to the complexity of the system and to the high loading, chlorine levels should be the same as for freshwater systems.

Several recommendations can be made regarding chlorination processes in power plants. First, intermittent, low level chlorination is the preferred technique. Continuous chlorination should not be used since there is substantial damage to entrained plankton even at very low chlorine dosages.

Past experience with power plants indicates that chlorination at higher than required concentrations has occurred at many sites. Many companies have made great efforts to reduce the duration of chlorination and concentrations of chlorine used. Fine tuning of the power plant to the particular cooling water system appears to be profitable.

Adequate control techniques to ensure low level chlorination are necessary. Chlorine gas is perhaps a more feasible control technique than either sodium or calcium hypochlorite because chlorine gas can be metered more accurately than the dry chemical. In many plants, it may be appropriate to alternate chlorination application so that only sections of the intake or condenser systems (split condenser chlorination) are chlorinated at a given time. This would certainly minimize both mortalities in the entrainment water and initial and latent mortalities of organisms in the receiving waters.

During periods of limited growth of fouling organisms, cessation of chlorination is recommended. The use of mechanical scrubbing systems is encouraged, particularly in new plants. Since little is known regarding the organic chlorinated compounds formed as a result of chlorinating and there is some concern over the carcinogenicity of chlorinated compounds, we recommend that alternative biocidal techniques be investigated (for example, see Yu et al. 1977).

It seems logical in biological terms to increase the biocidal effectiveness by having a higher ΔT in the plant. Higher volumes, although increasing the mechanical damage of entrained organisms, may reduce the number and types of fouling organisms, thus reducing chlorine usage. By entraining less water, mechanical effects are reduced; however, low volumes tend to increase the amounts of chlorine used. Lower flow volumes coupled with even higher ΔT's in combination with chlorination may be the best solution to minimizing entrainment losses. Obviously, the choice is: to allow organisms to survive plant passage or to sacrifice the entrained organisms. Depending on the site and the organism assemblage, a 100% kill, a 1% kill, or some intermediate kill of the entrained organisms may be preferable, depending on the community structure in the plant area.

For cooling tower systems, it is recommended that blowdown products not be released into the receiving waters, but instead be placed into a receiving pond prior to subsequent release into the receiving waters. If a power plant fitted with cooling towers is adjacent to shellfish areas, then no discharge of blowdown should be allowed. Based on dilution factors and site characteristics, some release of blowdown material may be allowed. These recommendations are based on Becker and Thatcher's

(1973) and Draley's (1972) information relating chemical usage in cooling tower system plus the assumption of a concentration factor of at least 2 to 3 times greater than ambient.

ACKNOWLEDGEMENTS

We thank Dr. Joel C. Goldman (Woods Hole Oceanographic Institution) and Mr. Thomas A. Miskumen (American Electric Power Service Corporation) for assistance in providing information. Mary Jane Reber assisted in typing and Martha Cole prepared several illustrations. This is Contribution No. 735 from the Chesapeake Biological Laboratory of the Center for Environmental and Estuarine Studies of the University of Maryland and Contribution No. 185 of the Marine Sciences Research Center of the State University of New York at Stony Brook.

REFERENCES CITED

Alderson, R. 1972. Effects of low concentrations of free chlorine on eggs and larvae of plaice, *Pleuronectes platessa L. In* Mar. Pollut. and Sealife. M. Ruivo (Ed). Fishing News (Books) Ltd. London. pp. 312-315.

Alderson, R. 1974. Seawater chlorination and the survival and growth of the early developmental stages of plaice, *Pleuronectes platessa* L. and Dover sole, *(Solea solea* L. Aquaculture. 4:41-53.

Arthur, J. W. and J. G. Eaton. 1971. Chloramine toxicity to the amphipod *Gammarus pseudolimnaeus* and the fathead minnow *(Pimphales promelas)*. J. Fish. Res. Board Can. 28:1841-1845.

Baker, R. and S. Cole. 1974. Residual chlorine: something new to worry about. Indust. Wat. Engineer. 11:10-20.

Basch, R. E. and J. G. Truchan. 1974. Calculated residual chlorine concentrations safe for fish. Michigan Water Res. Commission Report. (Unpublished, 29 pp.).

Becker, C. D. and T. O. Thatcher. 1973. Toxicity of power plant chemicals to aquatic life. Atomic Energy Comm. WASH-1249, UC-11.

Brook, A. J. and A. L. Baker. 1972. Chlorination at power plants: impact on phytoplankton productivity. Science. 176:1414-1415.

Brooks, A. S. and G. L. Seegert. 1977. The effects of intermittent chlorination on the biota of Lake Michigan. Center for Great Lakes Studies, Spec. Rept. 31. Milwaukee, Wis. 167 pp.

Brungs, W. A. 1973. Effects of residual chlorine on aquatic life. J. Wat. Pollut. Control Fed. 45:2180-2193.

Brungs, W. A. 1976. Effects of wastewater and cooling water chlorination on aquatic life. EPA Ecological Research Series, EPA-600/3-76-098, 46 pp.

Campbell, S. Y. and W. C. Runger. 1955. Use of chloride dioxide for algae control in Philadelphia. J. Am. Water Works. Assoc. 47:740-746.

Capuzzo, J. M. (in press). The effects of free chlorine and chloramine on growth rates and respiration rates of larval lobsters *(Homarus americanus)*.

Capuzzo, J. M., S. A. Lawrence and J. A. Davidson (in press a). Combined toxicity of free chlorine, chloramine and temperature to stage I of the American lobster *Homarus americanus*. Water Res.

Capuzzo, J. M., J. A. Davidson, S. A. Lawrence, and M. Libni (in press, b). The differential effects of free and combined chlorine on juvenile marine fish. Est. Coastal Mar. Sci.

Carpenter, E. J., B. B. Peck and S. J. Anderson. 1972. Cooling water chlorination and productivity of entrained phytoplankton. Mar. Biol. 16:37-40.

Cherry, D. S., S. R. Larrick, K. L. Dickson and J. Cairns, Jr. 1977. Continuation of laboratory determined preference and avoidance responses of New River and Lake Michigan fish to thermal and chlorinated discharges. Indiana and Michigan Elect. Co. Tanners Creek Plant. Chlorine Study Report Vol. II. Progress Rept. 67 pp.

Clark, J. and W. Brownell. 1973. Electric power plants in coastal zone: Environmental issues. Amer. Littoral Soc. Spec. Publ. No. 7.

Davies, R. M. and L. D. Jansen. 1975. Zooplankton entrainment at three mid-Atlantic power plants. J. Wat. Pollut. Control Fed. 47:2130-2142.

Dickson, K. L., A. C. Hendricks, J. S. Crossman and J. Cairns, Jr. 1974. Effects of intermittently chlorinated cooling tower blowdown on fish and invertebrates. Envir. Sci. Tech. 8: 845-849.

Dowty, B., D. Carlisle, J. L. Laseter and J. Storer. 1975. Halogenated hydrocarbons in New Orleans drinking water and blood plasma. Science. 87:75-77.

Draley, J. E. 1972. The treatment of cooling waters with chlorine. Argonne Nat. Lab., Ill. Env. State. Proj. ANL/ES-12, 11 pp.

Dressel, D. M. 1971. The effects of thermal shock and chlorine on the estuarine copepod, *Acartia tonsa*. M. S. Thesis, Univ. Virginia. 57 pp.

Eaton, J. W., C. F. Kolpin, H. S. Swofford, C-M. Kjellstrand and H. S. Jacob. 1973. Chlorinated urban water: A cause of dialysis-induced hemolytic anemia. Science 181:463-464.

Eppley, R.W., E. H. Ringer and P. M. Williams. 1976. Chlorine reactions with seawater constituents and the inhibition of photosynthesis of natural marine phytoplankton. Est. Coast. Mar. Sci. 4:147-161.

Eren, Y. and Y. Langer. 1973. The effects of chlorination on *Tilapia* fish. Bamidgeh. 25:56-60.

Eyman, L. D., C. W. Gehrs and J. J. Beauchamp. 1975. Sublethal effect of 5-chlorouracil on carp *(Cyprinus carpio)* larvae. J. Fish. Res. Board Can. 32:2227-2229.

Farkas, L. and M. Lewin. 1947. Estimation of bromide in presence of chloride. Anal. Chem. 19:665-666.

Fava, J. A. and C. Tasi. 1976. Immediate behavioral reactions of blacknose dace, *Rhinichthys atratulus,* to domestic sewage and its toxic constituents. Trans. Amer. Fish. Soc. 105: 430-441.

Fox, J. L. and M. S. Moyer. 1975. Effect of power plant chlorination on estuarine productivity. Ches. Sci. 16:66-68.

Gehrs, C. W., L. D. Eyman, R. J. Jolley and J. E. Thompson. 1974. Effects of soluble chlorine containing organics on aquatic environments. Nature. 24:675.

Gentile, J. H., J. Cardin, M. Johnson and S. Sosnowski. 1976. Power plants, chlorine and estuaries. EPA Ecological Research Series, EPA-600/3-76-055. 28 pp.

Ginn, T. C., W. T. Waller and G. J. Lauer. 1974. The effects of power plant condenser cooling water entrainment on the amphipod, *Gammarus* spp. Water Res. 8:937.

Ginn, T. C. and J. M. O'Connor (in press). Response of the estuarine amphipod *Gammarus daiberi* to chlorinated power plant effluent. Estuarine Coastal Mar. Sci.

Hamilton, Jr., D. H., D. A. Flemer, C. W. Keefe and J. A. Mihursky. 1970. Power plants: effects of chlorination on estuarine primary production. Science. 169:197-198.

Heath, A. G. Toxicity of intermittent chlorination to freshwater fish: influence of temperature and chlorine form. Indiana and Michigan Elect. Co. Tanners Creek Plant. Chlorine Study Report Vol. II, Appendix B, 28 pp.

Heinle, D. R. 1976. Effects of passage through power plant cooling systems on estuarine copepods. Environ. Pollut. 11: 39-58.

Heinle, D. R. and M. S. Beaven. 1977. Effects of chlorine on the copepod *(Acartia tonsa)*. Ches. Sci. 18:140.

Hirayama, K. and R. Hirano. 1970. Influences of high temperature and residual chlorine on marine phytoplankton. Mar. Biol. 7: 205-213.

Johannesson, J.K. 1958. The determination of monobromine and monochloramine in water. Analyst. 83:155-159.

Jolley, R. L. 1975. Chlorine containing organic constituents in chlorinated effluents. J. Water Pollut. Control Fed. 47: 601-618.

Kott, Y. 1969. Effects of halogens on algae. III. Field experiment. Water Res. 3:265-271.

Lanza, G. R., G. J. Lauer, T.C. Ginn, P. C. Storm and L. Zubarik. 1975. Biological effects of simulated discharge plume entrainment at Indian Point nuclear power station, Hudson River estuary, USA. pp. 45-123. *In* Combined effects of radioactive, chemical and thermal releases to the environment. International Atomic Energy Agency, Vienna.

Latimer, D. L., A. S. Brooks and A. M. Beeton. 1975. Toxicity of 30-minute exposures of residual chlorine to the copepods *Limnocalanus macrurus* and *Cyclops bicuspidatus thomasi*. J. Fish. Res. Board Can. 32: 2495-2501.

Lauer, G. J., W. T. Waller, D. W. Bath, W. Meeks, R. Heffner, T. Ginn, L. Zubarik, P. Bibko and P. C. Storm. 1974. Entrainment studies on Hudson River organisms, pp. 37-82. *In* L. D. Jensen (Ed.). Entrainment Workshop, Electric Power Res. Inst., Palo Alto, California.

Macalady, D. L., J. H. Carpenter and C. A. Moore. 1977. Sunlight-induced bromate formation in chlorinated seawater. Science 195: 1335-1338.

Mangun, L. B. 1929. Algae control by chlorination at Kansas City, Kansas. J. Am. Water Works Assoc. 21:44-49.

Marcy, Jr., B. C. 1973. Vulnerability and survival of young Connecticut River fish entrained at a nuclear power plant. J. Fish. Res. Board Can. 30:1195-1203.

Marcy, Jr., B. C. 1975. Entrainment of organisms at power plants with emphasis on fishes - an overview. pp. 89-118. *In* S. B. Saila, Fisheries and energy production: a symposium. Lexington Books, D. C. Heath Co., Lexington, Massachusetts.

Marcy, Jr., B. C. 1976. Planktonic fish eggs and larvae of the lower Connecticut River and the effects of the Connecticut Yankee plant including entrainment. pp. 115-139. *In* D. Merriman and L. M. Thorpe (Ed.). The Connecticut River ecological study. Monogr. No. 1, Amer. Fish. Soc.

Marinenko, G., R. J. Huggett and D. G. Friend. 1976. An instrument with internal calibration for monitoring chlorine residuals in natural waters. J. Fish. Res. Board Can. 33: 822-826.

Marks, H.C. 1972. Residual chlorine analysis in water and wastewater. *In* Water and Water Pollution Handbook, Vol. 3. Ciaccio, L. L. (Ed.). Marcel Dekker Inc., New York. pp. 1213-1247.

Mattice, J. S. and H. E. Zittel. 1976. Site-specific evaluation of power plant chlorination. J. Water Pollut. Control Fed. 48:2284-2308.

McLean, R. I. 1973. Chlorine and temperature stress on estuarine invertebrates. J. Wat. Pollut. Control Fed. 45:837-841.

Meldrim, J. W., J. J. Gift and B. R. Petrosky. 1974. The effect of temperature and chemical pollutants on the behavior of several estuarine organisms. Ichthyological Assoc., Inc. Bull. No. 11, 129 pp.

Merkens, J. C. 1958. Studies on the toxicity of chlorine and chloramines in the rainbow trout. Wat. Waste Treatment J. 7:150-151.

Middaugh, D. P., J. A. Couch and A. M. Crane. 1977. Response of early life history stages of the striped bass, *Morone saxatilis*, to chlorination. Ches. Sci. 18:141-153.

Mihursky, J. A. 1969. Patuxent Thermal Studies. NRI Special Report No. 1, Univ. of Maryland. 20 pp.

Morgan, II, R. P. and R. G. Stross. 1969. Destruction of phytoplankton in the cooling water supply of a steam electric station. Ches. Sci. 10:165-171.

Morgan, II, R. P. and R. D. Prince. 1977. Chlorine toxicity to eggs and larvae of five Chesapeake Bay fishes. Trans. Amer. Fish. Soc. 106:380-385.

Morris, J. C. 1967. Kinetics of reactions between aqueous chlorine and nitrogen compounds. pp. 23-53. *In* S. D. Faust and J. V. Hunter, Principles and Application of Water Chemistry. John Wiley and Sons, New York.

Muchmore, D. and D. Epel. 1973. The effects of chlorination of wastewater on fertilization in some marine invertebrates. Mar. Biol. 19:93-95.

Page, T., R. H. Harris and S. S. Epstein. 1976. Drinking water and cancer mortality in Louisiana. Science 193:55-57.

Podoliak, H. A. 1974. A review of the literature on the use of calcium hypochlorite in fisheries. U.S. Fish. Wildlf. Serv. National Technical Information Service, Springfield, Virginia. PB-235. 444.

Roberts, Jr., M. H., R. Diaz, M. E. Bender and R. Huggett. 1975. Acute toxicity of free chlorine to selected estuarine species. J. Fish. Res. Board Can. 32:2525-2528.

Roberts, Jr., M. H. 1977. Bioassay procedures for marine phytoplankton with special reference to chlorine. Ches. Sci. 18:130-136.

Sawyer, C. N. and D. L. McCarthy. 1969. Chemistry for sanitary engineers. 2nd Ed. McGraw-Hill, New York.

Seegert, G. L., H. S. Brooks and D. L. Latimer. 1977. The effects of a 30-minute exposure of selected Lake Michigan fishes and invertebrates to residual chlorine. pp. 91-99. *In* L. D. Jensen (ed.) Biofouling Control Procedures. Marcel Dekker, Inc., New York.

Shih, L. K. and J. Lederberg. 1976. Chloramine mutagensis in *Bacillus subtilis*. Science 192:1141-1143.

Sprague, J. B. and D. E. Drury. 1969. Avoidance reactions of salmonid fish to representative pollutants. Vol 1. Proc. 4th Int. Conf. Wat. Pollut. Res. Prague. 1969:169-179.

Stalham, C. N., M. J. Melancon, Jr., and J. J. Lech. 1976. Biocentration of xenobiotics in trout bile: A proposed monitoring aid for some waterborne chemicals. Science 193: 680-681.

Stober, Q. J. and C. H. Hanson. 1974. Toxicity of chlorine and heat to pink (*Oncorhynchus gorbuscha*) and chinook salmon (*O. tshawytscha*). Trans. Am. Fish. Soc. 103:569-576.

Tompkins, J. A. and C. Sai. 1976. Survival time and lethal exposure time for blacknose dace exposed to free chlorine and chloramines. Trans. Amer. Fish. Soc. 105:313-321.

Truchan, J. G. 1977. Toxicity of residual chlorine to freshwater fish: Michigan's experience. pp. 79-89. *In* L. D. Jensen (Ed.) Biofouling control procedures. Marcel Dekker, Inc., New York.

Tsai, C. 1975. Effects of sewage treatment plant effluents on fish: A review of literature. Chesapeake Research Consortium, Inc. Publ. No. 36, University of Maryland CEES Contribution No. 637, 229 pp.

Walker, C. R. 1976. Polychlorinated biphenyl compounds (PCB's) and fishery resources. Fisheries. 1:19-25.

Waugh, G. D. 1964. Observations on the effects of chlorine on the larvae of oysters (*Ostrea edulis* L.) and barnacles (*Elminius modestus* Darwin). Anal. Appl. Biol. 54:423-440.

White, G. C. 1972. Handbook of chlorination. VanNostrand Reinhold Co., New York. 744 pp.

Yu, H. H. S., G. A. Richardson and W. H. Hedley. 1977. Alternatives to chlorination for control of condenser tube biofouling. EPA Report 600/7/77-030.

Zillich, J. A. 1972. Toxicity of combined chlorine residuals to freshwater fish. J. Wat. Pollut. Control Fed. 44:212-220.

CHAPTER 4. EFFECTS AND IMPACTS OF PHYSICAL STRESS ON ENTRAINED ORGANISMS

BARTON C. MARCY, JR.
ALLAN D. BECK
ROBERT E. ULANOWICZ

TABLE OF CONTENTS

I. INTRODUCTION

Environmental studies of power plants have recently shifted their emphasis from examination of the effects of heated discharges to studies of the impacts of entire cooling systems. One of the major impacts arises when planktonic organisms are carried into and through a plant with the cooling water. Because of their relatively immobile, free-floating character, planktonic organisms are highly vulnerable to being "entrained" or passively drawn into the cooling water condenser systems of power plants. More than 70% of estuarine animals have planktonic eggs and larvae. The environmental impact of entrainment is related to the composition and abundance of affected organisms, the numbers of organisms in the adjacent waters, survival rates during entrainment as related to natural survival, the ecological roles of entrained organisms, and their reproductive strategies. Abiotic factors affecting entrainment impact include the location of the power plant, the design and operation of the cooling system, the quantity of water withdrawn, and the ambient conditions of water used for cooling.

Plankton, small fishes, and invertebrates which pass through the intake screens intact are subjected to various and simultaneous stresses which often lead to inner-plant mortality (see Chapter 5). Power plants can be represented as large predators that not only reduce the abundance of vulnerable organisms but may also disrupt the community structure through selective mortality and enhancement of the growth of some of the surviving species. Entrainment survival is determined by:

- sizes, life-stages, and relative susceptibility to injury of the species involved,
- ambient temperatures and the quality of the withdrawn and receiving water,
- amplitude of the temperature rise (ΔT) as the water passes through the condenser cooling system,
- duration of exposure to elevated temperatures,
- pressure changes resulting from turbulence (shear forces) and acceleration, as well as physical abrasion during passage through the system,
- exposure to biocides used for fouling control, and
- gas bubble disease (the formation of air embolisms) possibly caused by pressure and temperature changes in the cooling system.

The extent to which physical damage contributes to the total inner-plant mortality has not been adequately assessed at most power plants. Certain entrainment studies are now beginning to separate the effects of physical, thermal, and chemical stresses and to describe their synergistic effects. Thermal and chemical stresses vary in length and magnitude, depending on thermal exposure regimes and chlorination procedures. Physical stresses, on the other hand, are continuously applied whenever cooling water is being pumped.

In passing through a power plant cooling water system, entrained biota experience an array of stresses. Fig. 1 illustrates the areas of stresses in a typical steam electric generating station with once-through continuous flow of cooling water.

A plant normally interfaces the receiving water body at the intake and discharge structures. Water and organisms enter the intake through heavy steel trash bars which block floating debris. Typical intake approach velocities are on the order of 15 to 60 cm/sec; EPA (1973) recommends velocities of 15 cm/sec or less. The water then passes through a traveling screen with openings about half the diameter of the steam condenser tubes, or 0.75 to 1.25 cm in most cases. On reaching the pumps, organisms are exposed to pressure fluctuations, velocity shear forces, and physical buffeting and abrasion. The water passes into a large water box where velocities may increase up to eight times, then through two right angle turns, through banks of several thousand 2.3 cm (I.D.) condenser tubes where heat exchange takes place, and down and out a discharge pipe or canal to the receiving water body.

Entrained organisms are stressed by mechanical buffeting, acceleration, velocity shear forces, and changes in hydrostatic pressure. Inner-plant physical stresses arise from:

- contact with fixed or moving equipment, such as screens, pumps, and piping,
- pressure changes, especially the negative pressures or vacuums within the pumps,
- shear forces in areas of extreme turbulence or boundary proximity,
- accelerative forces resulting from changes in velocity and direction,
- buffeting and collision with the particles (i.e., organic load) passing along with the organisms, damage depending on load density and size, and
- cavitation in regimes of partial vacuum.

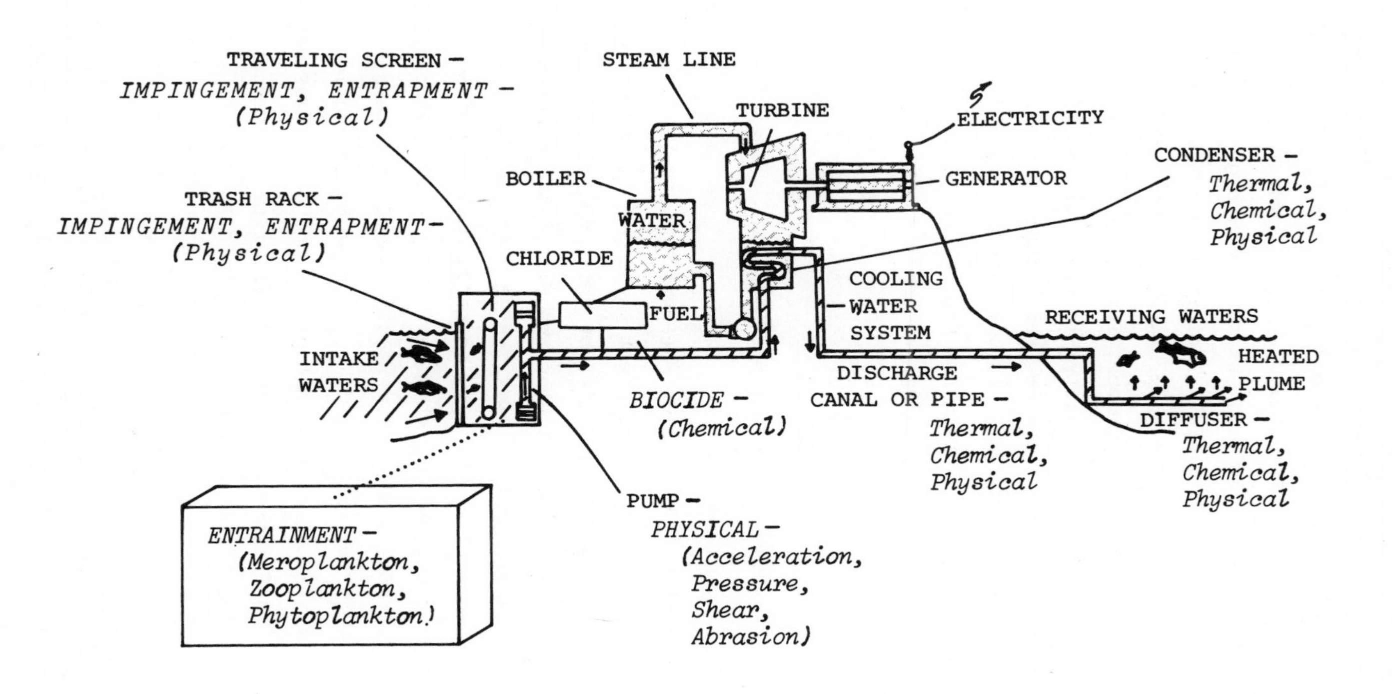

STEAM ELECTRIC GENERATING STATION
TRAVELING SCREEN – IMPINGEMENT, ENTRAPMENT – (Physical)
TRASH RACK – IMPINGEMENT, ENTRAPMENT – (Physical)
INTAKE WATERS
ENTRAINMENT – (Meroplankton, Zooplankton, Phytoplankton)
STEAM LINE
BOILER
WATER
CHLORIDE
FUEL
TURBINE
ELECTRICITY
GENERATOR
CONDENSER – Thermal, Chemical, Physical
COOLING WATER SYSTEM
BIOCIDE – (Chemical)
PUMP – PHYSICAL – (Acceleration, Pressure, Shear, Abrasion)
DISCHARGE CANAL OR PIPE – Thermal, Chemical, Physical
RECEIVING WATERS
HEATED PLUME
DIFFUSER – Thermal, Chemical, Physical

Fish culturists have recognized the impact of physical trauma on developing fishes for many years (Hayes, 1949; Davis, 1953; Leitritz, 1963). Physical disturbance during early development causes many of the deformities observed, e.g., misshapen heads in carp fry (Mätlak, 1970), deformed yolk sacs in alevins (Emadi, 1973), and vertebral compression (Mathur and Yazdani, 1969). After eye pigment develops, larvae of white flounder are extremely sensitive (50% mortality) to physical transfer shock for approximately 5 days (MacPhee, pers. comm.). Despite evidence of trauma due to relatively mild physical stresses, until recently little attention has been given to the physical rigors experienced by an organism passing through a cooling system.

II. FIELD STUDIES SHOWING PHYSICAL DAMAGE

Eraslan et al. (1976) point out that entrainable life stages could be exposed to vastly different physical stresses at different power plants. The mortalities of entrained species observed in field studies at 15 operating power plants are summarized in Table 1, Chapter 5. Table 1 provides data on mortality caused by physical damage alone as compared to total mortality from the combined effects of thermal, chemical, and physical stresses. Percentages of total mortality caused by physical stress as reported in the literature are in many cases approximations and are often difficult to compare because of the use of varied methodologies in field collection and mortality determination techniques.

A. Fish/Ichthyoplankton

Marcy (1975) and Adams (1968) indicate that physical stresses may have a much greater impact on fish eggs and larvae than does temperature. The limited data that are available

Table 1 Summary of Percentage Physical Damage Mortality of the Total Mortality Observed in Plankton Entrainment Studies at Power Plants

Power Plant	Reference	Percent Physical Damage	Percent Total Mortality
	Fish/Ichthyoplankton		
Millstone Point	Nawrocki (1977)	22.8	22.8
Chalk Point	Morgan et al. (1973, 1976)	20-50	-[1]
Calvert Cliffs	Morgan et al. (1973, 1976)	50-100	-[1]
Connecticut Yankee	Marcy (1971, 1973)	80	100
Ludington Pump Storage	Serchuk (1976)	37.2-61.5	56.5-67.7
Vienna	Flemer et al. (1971a)	99.7	99.7
Brayton Point	EPA (1972a)	100	100
Northport	Austin et al. (1973)	27-57	27-57
Indian Point	Lauer et al. (1973)	7-39	-[1]
Brunswick	Copeland et al. (1975)	30	0-50
Monticello	Knutson et al. (1976)	8.5-42.4	100
Nanticoke	Teleki (1976)	49.5	96.3-99
	Zooplankton		
Prairie Island	Middlebrook (1975)	20.5	20.5
Big Rock	Grosse Ile Lab. (1972)	29.5	-[1]
Waukegan	Industrial Bio-Test Lab. (1972)	6.0[2]	-[1]
Zion	McNaught (1972)	Some[2]	-[1]
Point Beach	Edsall and Yocum (1972)	8-19[2]	-[1]
Connecticut Yankee	Massengill (1976)	<1	100
Millstone Point	Carpenter et al. (1974)	70	100
Morgantown	Beck and Miller (1974)	40	60
Palisades	Consumers Power (1972a)	5-7	-[1]
Point Beach	University of Wisconsin (1972)	4-14	-[1]
	Phytoplankton		
Allen	Hardy (1971)	11	-[1]
Morgantown	Flemer et al. (1971b)	13	-[1]
Alamitos and Haynes	Briand (1975)	Some	42
Florida plants (4)	Weiss (Unpub. ms)	Some	70-90
Millstone Point	AEC (1974)	74	-[1]
Kewaunee	Wisconsin Public Service Corp. (1974)	11	-[1]
Palisades	Consumers Power Company (1973)	0-68	0-81
Zion	Restaino et al. (1975)	11-12	3-20

1 Unspecified
2 Highest percentage of total mortality

concerning the passage of fish eggs and larvae through power plants show mortalities between 28 and 100% (many near 100%) due to various combined stresses (Table 2). Mortality of entrained ichthyoplankton at the 16 operating power plants discussed below (including those in Table 2) ranged from 0 to 100% and averaged 72%.

Physical damage caused 80% of the 100% mortality of the young of nine fish species (2.6 to 40 mm) entrained at the Connecticut Yankee plant (Marcy, 1971, 1973). Almost all (99.7%) of the striped bass eggs which passed through the Vienna, Maryland generating station were killed and a high percentage of the eggs disintegrated (Flemer et al., 1971a). The 100% mortality of 164.5 million menhaden (5 to 50 mm) during one day at the Brayton Point Plant in Massachusetts was attributed to physical damage (EPA, 1972a). Passage through the Northport, New York plant caused 27 to 54% maceration mortality of four species of juvenile marine fish (Austin et al., 1973). Nawrocki (1977) found that damage to 17 larval marine fishes by physical stresses averaged 22.8% and that several species of clupeid larvae sustained 62.5% physical damage.

Schubel (1973) noted that ΔT's up to 10°C were not detrimental to the egg development or hatching success of five estuarine fish and that physical damage may have a much greater impact on development and hatching success than does temperature.

When juvenile and adult fish passed through the Ludington pump storage facility, 37.2 to 61.5% were physically damaged (of a total mortality between 56.5 and 67.7%) (Serchuk, 1976). Most of the damaged fish were lacerated or decapitated, indicating that physical contact and shearing forces caused the damage. At the Brunswick nuclear plant, with no heat added to the system, entrainment mortality of fish larvae ranged between 9 and 50%, about one third of which was attributed to physical damage (Copeland et al., 1975). Mortalities of the ten species of larvae entrained were highest during the summer. Fragile anchovies suffered the highest physical damage. Knutson et al. (1976) passed marked fathead minnows (30 to 60 mm) through the Monticello nuclear station and found that physical damage caused between 8.5 and 42.4% mortality while thermal damage caused only 8.6% mortality. At the Nanticoke generating station, Long Point Bay, Lake Erie, Canada, Teleki (1976) found 96.3 to 99% mortality of entrained young-of-the-year fish. Smelt *(Osmerus mordax)* constituted 95% the total larvae entrained. Physical injury accounted for 49.5% of the plant-induced mortality.

B. Zooplankton

In a recent review of the literature, the staff of the AEC (1973) found that losses of zooplankton in cooling systems ranged from 15 to 100% but suggest that 30% "may be more representative". Middlebrook (1975) states that studies of power plants (e.g., Icanberry and Adams, in press; Davis and Jensen, 1974) showed negligible to 15% mortality of zooplankton.

Mortality of entrained zooplankton at the 18 plants discussed below ranged from 0 to 100% and averaged 36%. Zooplankton mortalities of 29 to 55% at the Big Rock power plant on Lake Michigan were attributed to both physical and heat damage (Grosse Ile Laboratory, 1972). When EPA (1972b) compared the Big Rock plant and the Escanaba plant, where there was only a 7% mortality, they attributed the difference in mortality to higher physical damage at Big Rock because of its longer discharge pipe. According to McNaught (1972), physical damage caused the greatest mortality of zooplankton passing through the Zion plant. Data on passage of zooplankton through the condensers at the Waukegan plant show an average mortality of 6.0% due to physical effects and 1.8% due to thermal stress (Industrial Bio-Test Laboratories, Inc., 1972). Studies of zooplankton at the Point Beach nuclear plant show that physical damage (8 to 19% mortality) during condenser passage was more critical than thermal impact (Edsall and Yocum, 1972). Entrained zooplankton mortality attributed to only

TABLE 2 Fish Eggs and Young: Inner-plant Mortalities, Entrained, as Noted in

Source	Location	Percent Mortality	Species Entrained
Marcy (1971, 1973, 1976)	Connecticut Yankee Plant, Haddam, Connecticut River	100	Alewife, Blueback herring, White perch, Carp, White catfish, Spottail shiner, American shad, Johnny darter
Profitt (1969)	White River, Indiana	100	Spottail shiners
Clark and Brownell (1973)	Indian Point Plant, Hudson River, New York	97.5	Not specified
Carpenter et al. (1972)	Millstone Point Plant, Niantic, Connecticut	High	Flounder, Menhaden, Blueback herring
EPA (1972a)	Brayton Point Plant, Mount Hope Bay, Mass.	100	Menhaden, River herrings
Lauer et al. (1974)	Indian Point Plant, Hudson River, New York	39	Striped bass, Hogchoker, White perch, River herring
Flemer et al. (1971a)	Vienna Plant, Maryland	99.7	Striped bass
Flemer et al. (1971b)	Chalk Point Plant, Maryland	92.4	Not specified
AEC (1973)	Oyster Creek Plant, New Jersey	100	24 species

Estimated Numbers Lost, and Percentage of Population Entrainment Studies at Power Plants

Life Stage	*Estimated Numbers Lost*	*Percentage of Intake Water Body Population Entrained*
Larvae and juveniles 97.5% (2.6-15.0 mm) 2.5% (15.1-40 mm)	*179 million/ year*	*4.0% (1.7-5.7%) withdrawal of 61% of river larval population*
Young	--	--
Larvae and early juveniles	--	--
Larvae and juveniles	*Up to 20 million/day*	--
Larvae (5.0-50 mm)	*7-165.5 million/day*	--
5 species of eggs 17 species of larvae	--	--
Eggs	--	--
Larvae	--	--
Eggs and larvae	*150 million eggs/year 100 million larvae/year*	--

TABLE 2

Source	Location	Percent Mortality	Species Entrained
Nawrocki (1977)	Millstone Point Plant, Niantic, Connecticut	27.8	Clupeids, Tautog, Sea robin, 16 species total Clupeids-67.5% mortality Sea robin-32.7% mortality Tautog-44.7% mortality
Voightlander (1974)	Browns Ferry Nuclear Plant, Tennessee River/ Wheeler Reservoir	-	90-95% Gizzard or Threadfin shad
Carlson and McCann (1969)	Cornwall Pump Storage Plant, Hudson River, New York	-	Striped bass
Copeland et al. (1975)	Brunswick Nuclear Plant, Southport, N.C.	9-50	Anchovy, Goby, Croaker, Spot
Knutson et al. (1976)	Monticello Nuclear Plant, Mississippi River, Minn.	-	33 species
AEC (1974)	Millstone Point Plant, Niantic, Connecticut	Most Killed	13 species, (Menhaden, Blueback herring, Grubby, Winter flounder)
Hess et al. (1975)	Millstone Point Plant, Niantic, Connecticut	Most	Winter flounder
Edsall (1976)	Monroe Plant, Lake Erie	High	Not specified

(Continued)

Life Stage	Estimated Numbers Lost	Percentage of Intake Water Body Population Entrained
Larvae	--	--
Larvae	*1.17618 x 10^{10}/ 120 period*	*2.8% of 4.10208 x 10^{11} in reservoir (open cycle-0.91-2.81%) (closed cycle 0.5-0.14%)*
Eggs	--	*4%*
Larvae	--	*12% (Withdrawal from river cross-section)*
Larvae	*5 x 10^4 larvae/ pump day 3.6 x 10^6 larvae/pump day*	--
Larvae, Juveniles	*9.2 ± 4/hour 22,635/hour*	--
Larvae	*50 million Aug. 10-21*	*2.5 million winter flounder entrained (April-late June) of 1-6 million estimated in Niantic Bay*
Larvae	--	*1% reduction in Niantic Bay recruitment due to entrainment losses*
Larvae	*300 million April-Aug. 1973*	--

physical effects ranged from 2.5 to 14.9% at five Lake Michigan plants; physical effects alone accounted for most mortalities as compared to combined physical and thermal effects (NALCO Environmental Sciences, 1976). A 5 to 30% loss in mobility of zooplankton as the result of passage through just the pumps and condensers was observed at many plants (Statement by Dr. Wright, Westinghouse Environmental Systems Department, in testimony at the Wisconsin hearings, Department of Natural Resources, 1971). Less than 1% of the zooplankton which passed through the Connecticut Yankee plant were physically damaged, although nearly all insect pupae and a cladoceran were selectively killed by physical stress within the plant (Massengill, 1976).

Two recent studies show much higher mortalities of zooplankton from the physical component than were reported in early studies of power plants. Carpenter et al. (1974) report that 70% of the copepods entering the cooling water system of the Millstone Point plant at Niantic, Connecticut were killed by the physical or hydraulic stresses of passage and that total mortality ranged from 69 to 83%. Beck and Miller (1974) cite unpublished data which indicate that physical damage caused mortality of 50% of zooplankton and that the primary cause of the mortality was pumping effects.

In a two year study, more than 70% of the dominant inshore zooplankton species, *Oithona* spp. and *Acartia tonsa*, were killed at the Turkey Point plant on Biscayne Bay in Florida (Prager et al., 1971). Conditions that damaged the entrained plankton populations in the first year also prevailed in the second year, although dilution water which had passed through unheated condensers was added in the second year. "In the dilution water, whatever damage was due to hydrodynamic forces persisted, although no additional temperature stress was imposed by Unit 3. If temperature were the only stress upon the plankton, one would expect to encounter less than half the plankton mortality per unit volume of water than occurred in last year's samples since the dilution factor is about 2X......Although temperature increases in August of 1971 were less than half those the plant caused in 1970 the zooplankton mortalities were only 10 -20% less..." (Prager et al., 1971). Thus, physical damage at the Turkey Point plant may cause mortalities up to 50%.

Mysid and polychaete larvae (1.5 to 5.0 mm) suffered severe physical damage as the result of entrainment at the Morgantown nuclear plant in Maryland (Gentile and Lackie, unpubl. ms). Sandine (1973) showed that physical "abuse" in the cooling system injured larger macrozooplankters (arrow worms, ctenophores, and coelenterates). Mihursky and Dorsey (1973) indicate that large ctenophores suffered greater mortality from physical effects than small ones. Copeland et al. (1975) note little difference between percentage of mortality at the intake and discharge (2 to 43% intake vs. 2 to 49% discharge). Mortality of entrained lobster larvae at the Pilgrim nuclear plant in Massachusetts was due to physical action, heat, and chemicals (AEC, 1972).

Copepods entrained at the Turkey Point nuclear plant in Biscayne Bay, Florida were damaged by a combination of mechanical turbulence and chlorine (larvae survived to 40°C) (Lackey and Lackey, in press). Consumers Power Company (1972a, 1972b) and Benda (1972) note that zooplankton mortalities between 8 and 13% were caused by a combination of thermal and physical damage. At four fossil-fuel sites in Florida, zooplankton abundance was reduced by 70% at one plant and by 20% at three other plants, due in part to physical damage (Weiss, unpubl. ms). Coutant (1970) suggests that physical destruction during passage through a power plant caused the reduced number of zooplankton carcasses in the discharge canals. Physical effects produced little copepod mortality in summer but were a major factor during the cooler months at the Crystal River site in Florida (Maturo et al., 1974). Middlebrook (1975) found a statistically significant 20% mortality of zooplankton at the Prairie Island nuclear station after cooling tower entrainment; thermal elevation (ΔT) did not appear to be a significant cause of mortality. Conversely, at the Crane Power Plant on a tributary of Chesapeake Bay, Davies, Hanson, and Jensen (1976) found no significant increases in mortality because of physical effects.

C. Phytoplankton

There appears to be a conflict of results on the effects of phytoplankton entrainment. Few data are available concerning only physical effects on phytoplankton. Adverse impact has been demonstrated in several cases, but the contribution of physical stress to total mortality is not known. At two power stations, productivity was stimulated between 18 and 30%, while at six stations mortality ranged from 11 to 100% and averaged about 55%. Some stimulation of photosynthesis from the physical effects of condenser passage occurred at the Allen generating plant on Lake Wylie, North Carolina (Gurtz and Weiss, 1972). Williams (1971) found no significant physical effects and Smith and Brooks (1971) found that some stimuation (17.5 to 30%) in productivity could occur from physical effects. On the other hand, Hardy (1971) observed that factors other than heat reduced productivity by 11%. Flemer et al. (1971b) found a 13% reduction in productivity which they attributed to physical effects. Morgan and Stross (1969) concluded that physical damage, cell disruption due to chlorination, or some other factor inhibited photosynthesis in samples of the Chalk Point effluent.

Entrainment studies of marine phytoplankton passing through two southern California coastal plants indicate that plant passage disrupted the community severely by reducing diversity, promoting differential survival of some species, and reinforcing the dominance of the two major species (Briand, 1975). Mortalities, a portion of which were physically related, approached 42% during passage. As a result of his findings, Briand (1975) advocates using deep sea water rather than shore zone water for cooling.

Some inhibition of phytoplankton after condenser passage (ΔT 5.9°C) at the Michigan City plant was noted and photosynthesis was significantly decreased between intake and discharge (ΔT 6.9°C) at the Bailly plant (Arnold and the Pennsylvania Cooperative Fishery Research Unit, 1975). Industrial Bio-Test Laboratories, Inc. (1971) found no significant phytoplankton kill at the Waukegan plant.

Carbon fixation (primary production) was reduced by 98% at the Morgantown plant, due primarily to heat effects (Bongers et al., 1973). Gross photosynthesis was reduced 70% by condenser passage at four fossil-fueled plants in Florida and abundance (total cell numbers) was reduced 80 to 90%; reductions were attributed in part to physical abrasion (Weiss, unpubl. ms). Two years of studies at the Chesterfield plant on the James River in Virginia showed nearly 100% mortality of phytoplankton at times in the summer. These mortalities were probably due primarily to heat and may have contributed to critically depressed oxygen levels (Smith and Jensen, 1974). Passage of diatoms through the Millstone Point plant without chlorination or thermal addition resulted in mortalities as high as 72% for five species (AEC, 1974).

III. BIOTA-MORTALITY RELATIONSHIPS

The effects of plant passage on entrained organisms, especially physical impacts, can cause changes in community structure through changes in diversity caused by elimination of less tolerant species and life stages, and size selectivity because of damage or mortality to various life stages of species.

A review of the data and of Table 1, Chapter 5 provides insight into physical and biological factors related to mortality, including:

- organism size, shape, life stage mortality,

- o species-life stage susceptibility,
- o plant design and operational effects.

A. Size/Life Stage Related Mortality

Increased physical injury with increased size during plant passage has been reported for a few species of fish (Markowski, 1962; Oglesby and Allee, 1969; Marcy, 1971, 1973; Beck and Lackie, 1974). Marcy (1973), for example, found that the greatest physical damage occurred at night when larger fish (20 to 40 mm) were available for entrainment. During the day, the majority of entrained fish were less than 15 mm long.

At the Waukegan plant, zooplankton larger than 0.9 mm suffered 17% mortality while sizes smaller than 0.9 mm suffered only 4% mortality; zooplankton larger than 2.0 mm had 21% immobility while those around 0.4 mm had only 4% (Industrial Bio-Test Laboratories, Inc., 1971, 1972). Based on zooplankton data from the Waukegan, Zion, and Kewaunee stations, physical effects during entrainment were more detrimental to larger organisms (>0.95 mm) than to smaller organisms (<0.5 mm) and percent immotility varied directly with the size of the zooplankton species (NALCO Environmental Sciences, 1976). At the Morgantown nuclear plant in Maryland, mortality of zooplankton at sizes between 0.05 and 0.5 mm was low and mortality of *Acartia tonsa* in the 0.5 to 1.5 mm range was 18%, but both mysid and polychaete larvae (1.5 mm to 5.0 mm) suffered severe physical damage (Gentile and Lackie, unpubl. ms).

Physical "abuse" during entrainment injured larger macrozooplankters (e.g., arrow worms, ctenophores, and coelenterates) (Sandine, 1973). At the Morgantown station, Mihursky and Dorsey (1973) found that larger ctenophores suffered greater mortalities than smaller ones and they suggest that the internal diameter of the condenser tube was a limiting factor in survival. Maturo et al. (1974) found that the extent of physical damage to zooplankton was related to size of the organism at the Crystal River,

Florida plant where *Oithona* suffered little mortality but *Labidocera* was affected throughout the year. In later studies at this same plant, Alden, Maturo, and Ingram (1976) concluded that "The mortality caused by mechanical damage associated with passage through the power plant seems to depend, at least in part, on the size of the entrained organism." They again found that damage was lowest for the smallest zooplankton (*Oithona* sp.), highest for the largest (*Labidocera* sp.) and intermediate for intermediate sizes. McNaught (1972) found that mortality due to abrasion varied from 3.1% when small zooplankton were most abundant to 11% when larger zooplankton were most abundant.

A hypothetical size (and time) related mortality plot was presented by Beck and Lackie (1974). A binomial regression analysis (size vs. mortality), based on larval fish and zooplankton mortality and size data, shows that mortality increased with size; however, the observed mortality was due to the combined effects of thermal and physical stresses (Fig. 2). Generally, the highest mortality of entrained organisms from physical stress was found with the relatively large (>15 mm) fragile fish larvae.

The generalized model provided below, after some alteration of the available data to permit statistical analysis, provides a rough approximation of expected mortality to entrained organisms from physical stress as related to size. The data utilized are not truly comparable because studies were conducted at various sites and used different sampling methodologies. Results are also influenced by species-specific tolerances to physical stress and by plant design and operating characteristics. The equation is:

$$\text{Arc-Sine (mortality/100)} = 0.07571 + 0.04053\ \text{(size)}$$

The equation relates size and mortality as a linear function, based on an arc-sine transformation of the data, where $r = 0.92$ and $R^2 = 0.846$. The term on the left side of the equation is an angle measured in radians and expressed as a decimal. To estimate

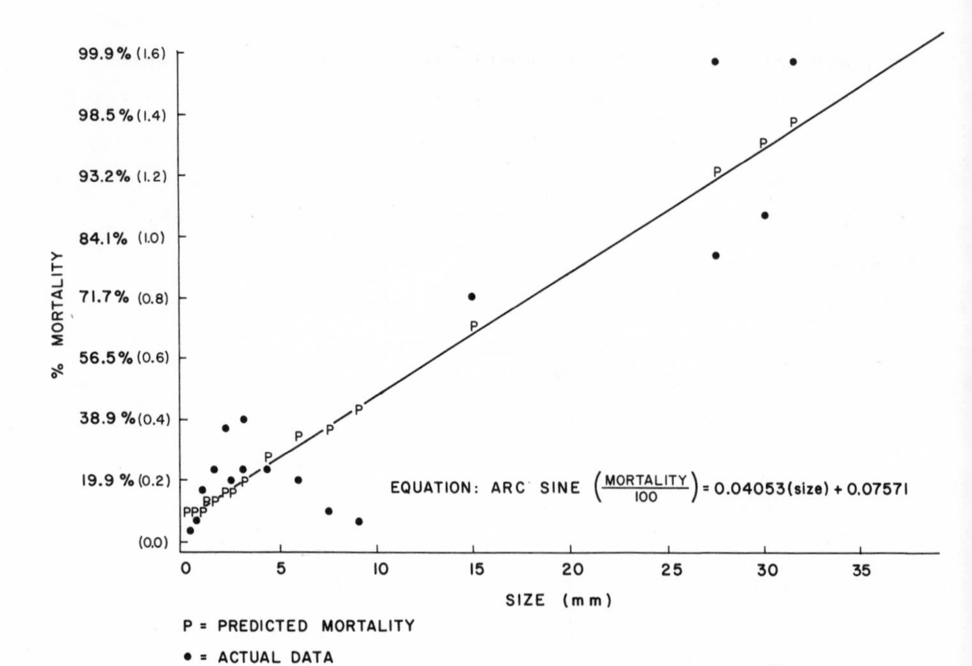

Fig. 2. Predicted size related mortality of ichthyoplankton and zooplankton in cooling water systems (Data from Marcy, 1975; Beck and Lackie, 1974; Nawrocki, 1977; EPA, 1972a and Tarzwell, 1972).

the mortality, the sine of the decimal number is multiplied by 100. For example, a 20 mm fish larvae yields an arc-sine (mortality/100) value of 0.88631 radians. The sine of 0.88631 is 0.7747, which multiplied by 100 results in an estimated mortality of 77.5%.

A few studies have showed mortality to decrease with increasing size. At the Millstone Point plant, mortality decreased from 36.3 to 9.3% as sizes increased from less than 2.25 mm to 10.25 mm (Nawrocki, 1977). At the Nanticoke Station on the Canadian side of Lake Erie, all smelt larvae longer than 14 mm were killed whereas only 77% of the smelt larvae 40 mm long were killed (Teleki, 1976). The results of the latter study support the observation that, once larvae are beyond a critical developmental period, they can withstand a greater degree of physical abuse; however, the results of the study are in contradiction to those of other studies and apply only to certain species, thus limiting the usefulness of the data.

B. Species Susceptibility

Size alone is only a partial measure of potential mortality of entrained organisms. Individual species in the same size range exhibit a wide range of susceptibility during plant passage. A review of Table 1, Chapter 5 reveals this wide variability in species mortality.

Different species and life-history stages show different susceptibilities to physical damage. The high percentage of physical damage (80%) noted in the Connecticut Yankee study was linked to the fact that 97.5% of the entrained fish were in the critical post-yolk sac stage and 97% of the larvae were fragile clupeids (Marcy, 1973). High physical damage of the fragile anchovy, as compared to nine other species entrained, was noted at the Brunswick nuclear plant (Copeland et al., 1975) and yolk and post-yolk sac stages were the most vulnerable to damage at the Millstone Point station (Nawrocki, 1977). Of the physical damage mortality of zooplankton entrained at the Connecticut Yankee plant, those species killed were considered fragile and had larger respiratory apparatus (Massengill, pers. comm.)

The head area, especially in fishes, appears to be the most vulnerable to physical damage (Marcy, 1971, 1973; Nawrocki, 1977; Knutson et al., 1976; Tarzwell, 1972). For example, of the mortality of marked fathead minnows (30 to 60 mm) passed through the Monticello plant's cooling system, 21% of the mortalities were due to decapitation, 63% were associated with damage to opercula-isthmuses-branchiostegal membranes, 2% with eye loss, and 14% with lateral cuts with bloody swellings (Knutson et al., 1976).

At the Millstone Point plant, where physical damage during entrainment to 17 species of larval fish ranged from 2.9 to 62.5%, larvae of the fragile clupeids, Atlantic silverside, sea robin, tautog, and cunner sustained the highest damages while those of the common pipefish experienced the lowest; eel larvae were unharmed (Nawrocki, 1977). In studies of plant passage at Northport, Long Island, maceration of fish larvae ranged from 27.6 to 57.1%, depending on the species (Austin et al., 1973). Menhaden larvae experienced 100% physical destruction (Tarzwell, 1972) at an estuarine plant in Massachusetts whereas no mortality of eel juveniles was found in studies at a Hudson River power generating station (Lauer et al., 1973).

Zooplankton mortality was selective at the Connecticut Yankee plant: all insect larvae and a predaceous cladoceran *(Leptodora kindtii)* were killed while other species were not (Massengill, 1976). Mortality of different species of copepods ranged from 7 to 100% in studies by Prager et al. (1970, 1971). Investigations at two west coast power plants report differential mortality to copepod species of the same genus: in a year-long study at Morro Bay, California, *Acartia tonsa* males exhibited 38.2% mortality whereas no mortality was observed for *A. longiremis* males. At Moss Landing, California, mortalities were

59.8, 27.8, and 23.2% for *A. tonsa, A. longiremis,* and *A. clausi* males, respectively (Icanberry, 1973). Heinle (1976) reported different sensitivities to inner plant passage of the copepods *Scottolana canadensis, Eurytemora affinis,* and *A. tonsa; S. canadensis* was the most sensitive, *A. tonsa* the least.

These data on zooplankton are general for combined stress effects of entrainment. It is probable that physical stress alone could cause similar species-specific mortality.

At Millstone Point, Connecticut, plant passage without chlorination or thermal addition resulted in differential kill of 17 species of phytoplankton (AEC, 1974). Mortality of five species of diatoms ranged up to 72% mortality while 12 other species were not significantly affected.

Briand (1975) found that entrainment effects on phytoplankton were very disruptive, changing the community structure and continually reducing species diversity. Passage through the condenser tubes affected algal species differently, killing diatoms in greater numbers (45.7%) than dinoflagellates (32.8%) and reinforcing the dominance of two major species. He indicated that only productive cells survived entrainment.

C. Plant Design and Operational Effects

Physical damage to menhaden larvae varied from 27.6 to 100% at two operating plants (Beck and Lackie, 1974). Mortality of *Acartia tonsa* males ranged from 0 to 9.09, 38.2, and 59.8% at four power plants on the west coast (Icanberry, 1973). Eraslan et al. (1976) point out that entrainable life stages could be exposed to vastly different physical stresses at different power plants.

Certain power stations have been reported to produce near total destruction of meroplankton while others cause much less damage (Jensen, 1977). Jensen postulates that these differences may be related to excessive turbulence and pressure from cavitation in inefficiently operated circulating pumps.

Davies, Hanson, and Jensen (1976) concluded that zooplankton mortality did not increase significantly because of physical stress at the Crane Power Plant, Chesapeake Bay. This contrasts with observed results at Millstone Point, Connecticut (Carpenter, Peck, and Anderson, 1974), Haddam Neck, Connecticut River (Marcy, 1974), and Morgantown, Maryland (unpublished data cited in Beck and Lackie, 1974).

Beck and Miller (1974) found that differences in effects of entrainment passage on organism mortality were generally different between west coast and east coast power plants at marine sites. There also appear to be plant-related differences in the effects of condenser passage on phytoplankton (Marcy, 1975).

Differential mortality of entrained organism as related to the design and operational characteristics of various plants is little understood at present. An understanding of the kinds and magnitudes of the entrainment stresses involved, coupled with a knowledge of the precise effects and sites of damage, should provide insight into the design of cooling systems which would minimize mortality of entrained species.

IV. STRESSES CAUSING PHYSICAL DAMAGE

A. Pressure

Little work has been directed toward the effects of the pressure changes encountered in power plants. The rapid pressure changes that occur in power plants may have the greatest potential for damaging entrained organisms, especially fishes (American Nuclear Society, 1974). Pressure changes encountered in plant passage may be sufficient to produce air embolism in fry. These embolisms, even if not lethal in themselves, may buoy fry to the surface, keep them in the warmest part of the plume, subject them to increased thermal shock, and cause them to become more vulnerable to predation (Edsall and Yocum, 1972). The following is an account of pressure differentials in a power plant cooling system:

> "Abrupt changes in pressure occur at various points in the circulating water system. As water approaches the impeller of the intake pump there is a rapid drop in pressure which is immediately followed by a pressure increase on the back side of the impeller. The magnitude of the pressure differential experienced by entrained organisms depends on the depth from which they and the cooling water are withdrawn and the design of the intake pump impeller. The positive pressure behind the intake pumps rapidly drops to low absolute values in the condenser system. The reduction to negative pressure is concurrent with the temperature rise and occurs within about 10 to 20 seconds during condenser passage. The maximum negative pressure is expected to occur at the condenser water box. As the flow enters the discharge system, there is a rapid return to positive pressure, the magnitude of which is dependent on the depth in the discharge system." (American Nuclear Society, 1974)

Negative pressures appear to be more damaging than positive pressures (American Nuclear Society, 1974). Changes in hydrostatic pressure should induce minimal physical strain upon an aquatic organism containing no gas vacuoles. Large positive pressure changes are usually benign unless the organism possesses a natural gas space, in which case the cavity may implode. Negative pressure changes, by contrast, have a high potential for inducing physical damage, especially during decompression. In organisms possessing a gas cavity, it may explode if the organism is unable to equilibrate the pressure across the membrane wall fast enough. Also, the solubility of dissolved gases drops as the ambient pressure drops. The water then becomes supersaturated with dissolved gases and, in the presence of living tissue, which serves as a form of catalyst, the gas may come out of solution and the consequent bubbles may cause physical trauma (gas bubble disease) (see Wolke et al., 1975).

Preliminary laboratory studies show that physical damage from the effects of pressure may be the largest cause of mortality to the larvae of bluegill, carp, and gizzard shad (Coutant, pers. comm.). Rapid pressure changes are characteristic of water passing through the cooling system: pressures fall just ahead of the pump impeller, build up to around 1.7 to 2.0 atm at the condenser tubes, and then fall to around 0.14 to 0.34 atm beyond the condenser tubes (Clark and Brownell, 1973). Pressure changes experienced by pump-entrained organisms as they pass from pumps to the discharge canal at the Indian Point plant ranged from 0.3 to 1.6 atm (Lauer et al., 1973). Roach (*Rutilus rutilus*) fingerlings exhibited 100% mortality at a pressure release rate of 3 atm/sec, 40 to 72% at 0.1 to 0.5 atm/sec, and 10% below 0.1 atm/sec (Tsvetkov et al., 1972). The high mortality of fish larvae passing through the Millstone Point nuclear plant was attributed, in large part, to pressure changes (2 atm at the water box dropping to 0.5 atm as the water falls from the condenser to sea level) (Nawrocki, 1977). Bioassays of the effect of pressure change on striped bass eggs and larvae showed little in the way of significant induced mortalities (Beck et al., 1975). Positive pressure impulses up to almost 100 atm yielded no strong pattern of mortality. Those mortalities judged to be nominally significant were largely associated with pressure drops from atmospheric to sub-atmospheric (0.3 atm) pressures or with organisms acclimated to higher pressures (because of water depth) and released to atmospheric pressures. Large pressure changes can also exist at plants with deep water intakes [New York University Medical Center (1975)]. However, pressure changes in problem areas can be easily manipulated and controlled during the design stages (O'Conner, pers. comm.). Studies which show pressure to have an impact on entrained organisms and which provide an overall review of the effects of changes in hydrostatic pressure on aquatic organisms include Beck et al. (1975),

Sleigh and MacDonald (1972), Zimmerman (1970), Kinne (1972), and Knight-Jones and Morgan (1966).

B. Acceleration

Acceleration forces, which result from changes in the velocities of flowing waters as they pass from the intake to the discharge area, may cause high mortality of entrained organisms (Ulanowicz, 1975). The lowest forces include accelerations due to changes in the bulk speed of water flow; they commonly range from very low to near that of gravitational force. These forces probably cause little damage. The next range includes those forces resulting from turbulent eddies. These forces are of such magnitude, usually several times that of gravity, as to possibly cause immediate or latent damage to entrained fish larvae. The highest range of acceleration forces is that of the short duration, high magnitude forces that result from impact with solid surfaces. Forces in this range are many times the force of gravity and, combined with damage from mechanical abrasion on impact, are probably immediately lethal. Only limited data are available on the effects of acceleration forces on aquatic organisms.

Post et al. (1973) examined the effects of acceleration rates 1, 2, 5 and 10 times that of gravity on different developmental stages of the relatively hardy rainbow trout and found that survival at hatching (72 to 84%) was not significantly different from that of the controls. Acceleration/shear stresses should have a stronger impact on more fragile organisms. Acceleration/shear forces of roughly 3 times that of gravity can be lethal to eggs and larvae of striped bass and white perch.

There are very few studies on the effects of acceleration on eggs and larvae, especially in terms of the forces experienced during plant passage. In two experiments for which the data are available, Battle (1944) and Rollefsen (1930) studied the effects of acceleration by dropping eggs of the cod, stickleback, mummichog, four-beard rockling from various heights on to a stretched piece of silk. Unfortunately, the heights of release (20, 40, and 60 cm) cannot be readily converted into impulse profiles. Using Newton's second law, the impulses per unit mass of egg work out to approximately 198, 280, and 343 dyne-sec per gram of egg mass, respectively, but the spring characteristics of the silk bed are not known and thus it is impossible to calculate the maximum inertial force associated with the impact. Speculating that the characteristic impact time is on the order of 50 ms would imply corresponding forces of 4.0, 5.7, and 7.0 times that of gravity. The experimental results indicate that these intermediate accelerating forces are potentially lethal to the fragile ichthyoplankton. Battle's (1944) results showed that vulnerability to physical stress among the four species tested varied widely and that damage decreased with increasing size and age of all four species tested.

C. Shear

Shear forces, expressed as units of force per unit area (dynes/cm^2) develop when spatial differences in velocity exist in a moving fluid, for example, at the edges of eddies in a turbulent flow regime or when water flows across a solid surface (Ulanowicz, 1975). The greatest shear stresses occur in close association with solid surfaces, such as pipe walls, pump impellers, traveling screens, and water boxes. Shear stress has two major components: rotation and deformation. When a fish egg is caught in a changing velocity field, the rotational effect of shear disturbs the internal order of the egg while the deformation effect stresses the outer membrane (Fig. 3), leading to possible break-up of the egg if the shear forces are high enough (Ulanowicz, 1975).

Morgan et al. (1973, 1976) attempted to measure the mortality caused by the shear fields that are created by water movement over the surface of fish eggs and larvae. Based on shear-time

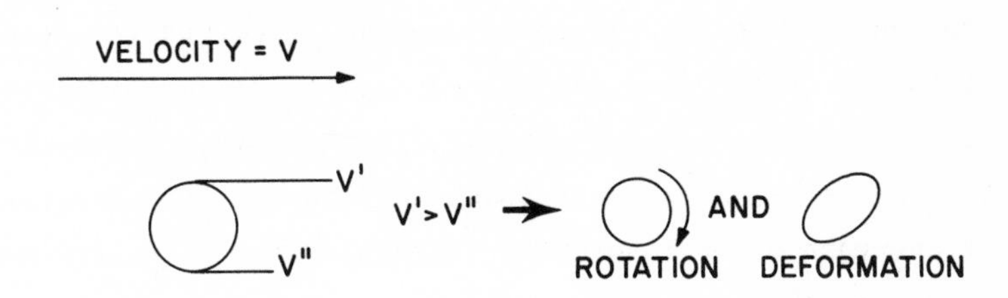

Fig. 3. Possible effects on a fish egg of shear stress resulting from water velocity.

exposure experiment, LD_{50} values were developed for white perch and striped bass. The results show that shear stresses between 120 and 785 dynes/cm^2 were lethal to 50% of the eggs and larvae at exposures of 1 to 20 min. The data obtained were used to predict the shear or physical stress created by water passing through the Chalk Point fossil-fuel plant. Preliminary results indicate conservative estimates of mortality between 20 and 50% for both species (Morgan and Ulanowicz, pers. comm.). Morgan calculated the shear force of water passing through the Calvert Cliffs nuclear plant and estimated that shear stress alone would be enough to cause 100% mortality of entrained eggs and larvae.

An interesting combination of field and laboratory data shows promise for estimating potential physical damage at a power plant, at least in terms of shear-caused damage. Three years of field studies at the Connecticut Yankee Atomic plant show that 80% of the 100% mortality of the young of nine fish species (2.6 to 40 mm) entrained was caused by physical damage (Marcy, 1973).

Clupeids made up 97.6% and white perch 1.3% of the total entrained species. The shear force of the waters passing through the Connecticut Yankee plant's intake, specifically at the walls of the water box, was calculated between 72 and 230 dynes/cm^2 (Ulanowicz, 1975; Morgan et al., 1976). Passage time from intake to the discharge canal at this plant is 93 sec. The LD_{50} shear stress values for laboratory tests on white perch and striped bass were 385 to 540 dynes/cm^2 (Morgan et al., 1976). According to these data, calculated shear forces in the water box are probably not lethal to white perch and striped bass. However, these shear forces may produce high mortalities of the fragile clupeids, which made up 97.6% of the entrained species, but no quantitative data are available on the ability of clupeids to withstand shear stresses.

It may be useful to attempt to relate the absence of physical impact observed by Coutant and Kedl (1975) with the shear bioassay experiments of Morgan et al. (1976). Multiplying the lethal shear (in dynes/cm^2) by the exposure time (in minutes) and averaging Morgan's short exposure trials for each organism, a lethal shear impulse of 570 and 833 dyne-min/cm^2 for striped bass eggs and larvae, respectively, and values of 635 and 900 dyne-min/cm^2 for the eggs and larvae of white perch are obtained. Assuming that the cross-sectional areas of eggs and larvae are characterized by diameters of 0.3 and 0.5 cm, respectively, and further assuming a density close to that of water, then the characteristic force on the organisms over a typical 1-min exposure period is estimated at 2.9 and 2.7 times that of gravity for striped bass eggs and larvae and 3.2 and 2.7 times that of gravity for white perch eggs and larvae.

If these assumptions are accepted, the experiments of Morgan et al. (1976) indicate that juvenile *Morone* spp. begin to exhibit mortality when the shear stress to which they are subjected approaches 3 times the force of gravity.

The condenser replica used by Coutant and Kedl (1975) was 1.89 cm (I.D.) wide and 1,220 cm long. When water and entrained larvae were passed through the tube under a 3 atm pressure drop, the mass-average velocity was 579 cm/sec. Using these figures, the characteristic shear stress at the wall was approximately 776 dynes/cm^2 and the exposure times were slightly over 2 sec. This converts into an impulse of 27.2 dyne-min/cm^2, which is well below those in the experiment of Morgan et al. (1976). Also, using the same assumptions given above on larval size and density, the characteristic force experienced by the test organisms is estimated at 2.4 times the force of gravity. Hence, the results of both Coutant and Kedl (1975) and Morgan et al. (1976) are consistent in indicating that the condenser tubes are an unlikely site for shear damage.

D. Abrasion/Collision

Abrasion can occur when two surfaces move in contact past one another or when a smaller suspended particle with a different velocity impinges upon an organism's surface. Abrasion is a difficult stress to quantify, since it is highly dependent upon the various natures of the contracting surfaces or colliding particles. It acts to decrease the lethal shear threshold of an organism. Under some circumstances it can become the limiting form of physical damage, as was the case in red blood cell destruction in early heart-lung machines. Although no quantitative data exist on abrasion as a factor in entrainment mortality, Emadi (1973) and Marcy (1976) mention abrasion as a possible factor in ichthyoplankton entrainment mortality.

A high silt and detritus load passing through the cooling system with the organisms may cause abrasion/collision mortality (Marcy, 1973, 1975, 1976; Coutant and Kedl, 1975). The average seston concentration in the Connecticut River near the Connecticut Yankee plant was 0.2306 g/m^3 and a concentration of 0.2338 g/m^3 passed through the plant during entrainment studies; such concentrations may be important in the physical damage mortality of certain organisms being pumped through the plant (Massengill, 1976). However, no data are available on hardness, size, shape, and angularity of the particles entrained.

A high organic load during plant passage may have caused the physical damage mortality of phytoplankton and zooplankton at four Florida fossil-fuel plants (Weiss, unpubl. ms). Entrained fish (2.6 mm to 18 mm) made up 2.5% of the total organic load passing through the Monticello nuclear power station. The high organic load possibly contributed to the high mortality observed (Knutson et al., 1976); unfortunately, no data on the characteristics of the particles are given.

V. AREAS PRODUCING PHYSICAL DAMAGE IN COOLING SYSTEMS

Areas where physical damage is likely to occur in a cooling system include fixed or moving equipment (e.g., piping and screens), circulating pumps, the water box, and condenser tubes.

A detailed description of a hypothetical organism's inner-plant passage is provided in Appendix A. The areas of physical stress in the system are discussed here as they relate to mortality of a diverse array of organisms as observed in field studies of operating electric power generating stations.

A. Pumps

Once the entrained organisms reach the pumps (Fig. 4), they are exposed to sudden pressure fluctuations, velocity shear forces, and physical buffeting and abrasion (Beck and Lackie, 1974). Once in the pump, rapid positive and negative changes in hydrostatic pressure, ranging from 0.29 to 1.6 atm, occur. In a few seconds, velocity shear forces fluctuate widely, along with severe buffeting and possible contact with pump walls or impeller blades. Physical damage is perhaps most severe in the pumps

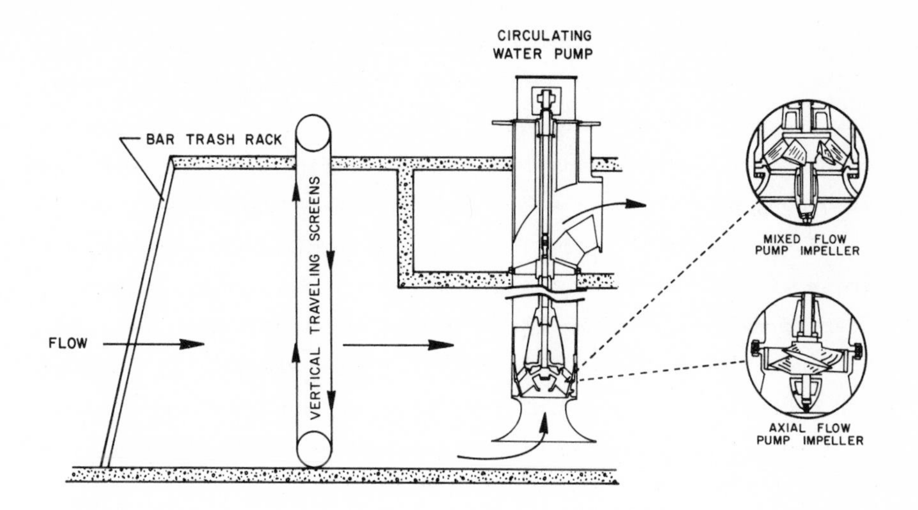

Fig. 4. Representative power plant cooling water intake forebay illustrating the internal characteristics of a typical circulating water pump and showing both axial-flow and mixed-flow impellers.

(Lauer et al., 1973). Gentile and Lackie (pers. comm.) demonstrated that the major cause of mortality to entrained phytoplankton and zooplankton was physical damage, due mainly to pumping effects. Their experiments showed that before the pumps the average mortality was 10%; after the pumps, 50%; and after condenser passage, 60%. In experiments concerned with passage of marked adult and juvenile fish (n = 2,742) through the Ludington pump storage plant's pumping system, mortality averaged 56.5 to 67.7% over a 2 yr period. Physical damage varied from 37.2 to 61.5% and most fish displayed lacerations or decapitation, implying that physical damage and shearing forces in the pump caused the damage (Serchuk, 1976).

The acceleration forces associated with the impact of an organism with an impeller blade are not easily estimated. Kedl (pers. comm.) reports a typical maximum velocity at the pump impeller tip to be about 1,860 cm/sec. Collision with such an object is equivalent to dropping the organism from a height of 17.7 m. In the light of Battle's (1944) results, mortality from such inertial forces is expected. Kedl (pers. comm.) has estimated that the probability of such a collision for a typical mixed-mode pump found in most power plants is about 2 to 5%.

Shear within the pump is an even more difficult quantity to assess with confidence. Using dimensional arguments, however, an order-of-magnitude estimate can be made as to the shear stress on the surface of the impeller blade. A typical power plant 350 hp pump operating at 73% efficiency dissipates approximately 7.31×10^{11} ergs/sec. This energy is passed across the impeller surface (ca. 1.15×10^4 cm^2), moving at a characteristic velocity of 1,860 cm/sec, across a high shear boundary layer into the turbulent flow field. Dividing the dissipated energy by the product of the impeller speed and surface area, a characteristic shear stress of 6,835 dynes/cm^2 at the impeller surface is obtained. This stress is roughly ten times the shear stress at the walls of the condenser tubes and is the highest shear stress and the most probable source of physical damage that an organism would experience during its passage through the cooling system.

Thus, both empirical and theoretical arguments point to the pump as the most likely site of physical damage and the section of the cooling system upon which those studying the impact of physical stress should concentrate their initial efforts.

B. Water Box

The water box can inflict physical damage because this area exhibits the maximum negative pressures found in the entire cooling system, negative pressures being the most damaging (American Nuclear Society, 1974). This area also has the highest flow rates (8 times higher than intake flows) (Marcy, 1973).

C. Condenser Tube

According to Coutant and Kedl (1975), the condenser tube is unlikely as the area where physical damage to fish larvae occurs. A single pass experiment of 2 wk old larval striped bass through a laboratory mock-up of a power plant condenser tube (not including a pump) resulted in mortalities no greater than that of the controls when only physical stresses were exerted. The experiments were conducted using different combinations of turbulent shear, pressure change, and temperature rise.

D. Cooling Towers

Most investigators and regulatory agencies have been assuming 100% mortality of plankton in closed cooling systems with cooling towers, mainly because of the "extreme temperature, mechanical and chemical stresses of the condenser cooling system" (NRC, 1976). Although much less water, and therefore less entrained organisms, is withdrawn from the receiving water body when cooling towers rather than once-through systems are used, few studies have been conducted to determine actual mortalities of plankton.

VI. SUMMARY AND CONCLUSIONS

Observations at operating power plants indicate that many species of organisms cannot tolerate passage through the cooling water system. A wide range of tolerances to plant passage (entrainment) is exhibited by various species. Generally, physical damage, principally occurring in passage through the pumps, is the major cause of mortality during the normal operational cycle of the plant. Thermal and chemical stresses can undergo temporal variations in magnitude of effects, depending on thermal exposure regimes and chlorination procedures. Physical stresses, however, are experienced continuously whenever cooling water is being pumped. Over 70% of estuarine animals have planktonic eggs or larvae (Beck and Lackie, 1974) and, based on present knowledge of biota-plant interactions, it appears that mortality of meroplankton and juvenile fishes due to passage through the plant is the paramount problem to be addressed in attempting to minimize potential environmental damage.

Available data indicate that mortality of entrained species is at least partially related to several factors: size and life stage, tolerance of the individual species and life stage, and differences in power plant cooling system designs and operational characteristics. At any one site, these factors are interrelated and the use of data to predict mortality, as related to specific characteristics of the biota and the plants' cooling system, is somewhat hampered by the complexity of the interrelationships involved. Also, the many field studies of operating plants are not truly comparable. Lauer et al. (1973) note, for example, that the literature is replete with inconsistent results. They suggest that problems in sampling design and methodology and the lack of understanding of the interaction of cooling systems and biota possibly lead to substantial interpretive error.

The impact of power plants can be minimized by siting in nonproductive areas and, in many cases, by utilizing closed cooling systems, or higher ΔT's, resulting in lower intake volumes, since the significance of the mortality of entrained organisms is directly related to the volume entering the intake. This low volume concept is presently the only immediately effective approach to minimizing adverse entrainment effects if high mortality is assumed or observed during plant passage. Also, it may be possible to increase condenser ΔT's while lowering intake volumes, especially when the physical damage component of the mortality dominates, as appears to the case with entrained meroplankton and juvenile fishes. Teleki (1976), for example, states that, at the Point Beach Lake Erie plant, fish loss due to entrainment would be reduced by decreasing the volume of cooling water. Most organisms were killed by physical or heat shock and reducing volume would expose fewer organisms to plant passage. He calculated that a 59% reduction of loss would be achieved by eliminating the tempering (augmentation) cooling.

At this time, there is a sufficient level of insight and understanding of biota-mortality relationships to support a gross level of predictive capability on entrainment mortality due to physical stress. It is hoped that further knowledge of biota-plant interactions will provide design criteria to maximize survival of organisms during passage through a plant where exclusion from the cooling system is not practical.

VII. RECOMMENDATIONS

A. Research

1. *Meroplankton studies should have highest priority.* Losses of meroplankton, including ichthyoplankton (fish eggs and larvae) and invertebrate larvae, to the aquatic system represent a larger impact on the ultimate population of adults over a broader geographical area than do losses of phytoplankton and zooplankton. This is because regeneration times or reproductives cycle of fish may require 2 to 4 yr as opposed to a few days for zooplankton and a few hours for phytoplankton. As a result, it is recommended that much more emphasis be placed on the effects on meroplankton, especially since many species are recreationally or commercially important.
2. *Cooling water pump warrants close scrutiny.* Available experimental evidence points to the pump as the major site of most of the physical damage inflicted on entrained organisms. Measured pressure values reveal that the greatest and most rapid pressure changes occur within the pump. Estimates based on dimensional considerations indicate that

the highest shear and acceleration forces are present near the impeller. Therefore, future field studies on physical damage should give special attention to possible mortalities within the pump. Further analysis of power plant pump design is needed to determine the magnitude and location within the apparatus of the greatest physical forces. The magnitude of these forces should be compared with the results of the physical stress bioassays. Stresses within alternate pump designs should be evaluated. Laboratory replicas of power plant pumps, properly scaled, could provide useful information on pump damage to various organisms.

3. *Bioassay data on physical stresses are urgently needed.* While some bioassay data exist for lethal doses of heat and biocides, very little information has been gathered on the intensities of shear and acceleration that entrainable organisms can endure. Such basic knowledge is imperative if the scattered work on physical damage is to be brought together within a strong conceptual framework. Research into the response of organisms to various levels of acceleration and shear for different exposure times, under precisely controlled laboratory conditions, is essential to our understanding of the problem of physical damage and is a prerequisite to the optimal design of cooling systems.
4. *Pressure bioassays require extension.* The bioassays on pressure are more complete and precise than most experiments on physical damage. Results to date indicate that pressure effects are not substantial in the operating range of most plants. There is no strong pattern apparent in the effects observed thus far, perhaps indicating that the threshold pressures that cause damage are just being approached in these experiments. It seems advisable to increase the range of applied pressure stress in these experiments to more fully delimit the lethal regions. The sections of the

cooling system most likely to cause pressure effects are the upstream side of the pump and the effluent plume.

5. *Abrasion deserves consideration.* Abrasion damage to entrained organisms may also be caused by collisions with particles of suspended materials passing through the plant with the organisms. The damage would be dependent upon such factors as the size and density of particles and their concentration in the cooling water. Few studies have estimated particle concentrations in the receiving and intake waters or have related such concentrations to degrees of physical damage. Such evaluations are recommended to assess abrasion damage during passage.

6. *Size and life stage susceptibilities need further definition.* Mortality appears to be directly related to size among organisms of similar susceptibility to physical damage. The larger the organism the greater the opportunity for contact with hard surfaces within pumps and piping systems and the greater the potential for damage by shear stress produced by discontinuities of water motion in turbulent flows.

   A tentative hypothesis relating the size of the organism to the rate of mortality was provided by Beck and Lackie (1974). A plot and equation derived from this hypothesis, when combined with additional field data, provide an estimate of mortality, based on the size of entrained organisms. There still are insufficient data and the relationship must be tested further with additional data to enhance its predictive value.

   Certain categories of organisms exhibit relatively high entrainment mortality. Large (total length 5 to 50 mm) clupeid and atherinid fish larvae are the most susceptible organisms. Relatively higher percentages of smaller organisms, including certain species of phytoplankton and zooplankton, survive plant passage.

Unfortunately, most of the data available are from different sites with different species and a wide range of study methodologies and personnel. Coutant (pers. comm.) and O'Connor (pers. comm.) have indicated that studies are being undertaken by the NRC at Oak Ridge, Tennessee and by the New York University Medical Center to simulate in the laboratory the physical stresses generally experienced by entrained organisms in passage through pumps and condensers. It is hoped that these types of investigations will provide the data needed to support or deny the tentative size/mortality hypothesis. Studies are needed on the various life stages and range of sizes of ecologically and/or commercially important representative species that are potentially entrainable. From such studies, the species most sensitive to physical stress would be identified, leading to a better understanding and predictability of entrainment effects on selected important species at individual power plant sites.

7. *Species vulnerability differences should be defined.* There may be a general relationship between size and mortality; however, species susceptibility among organisms of similar size is highly variable and appears to be a more important overall consideration than size alone. At one site, phytoplankton survival has been shown to vary from 11.4 to 87.6%, depending on species, with diatoms generally killed in larger numbers than dinoflagellates (Briand, 1975). Individual mortality by zooplankton in studies by Prager et al. (1971) ranged from 7 to 100%, depending on species. Nawrocki (1977) found that damage to fish larvae in the size range of 10 to 40 mm varied from 0 to 67.5%. Species-specific variability to physical stress strongly demonstrates the need for establishing tolerance data for a diverse array of entrainable species at all trophic levels.

B. Sampling

1. *Inter-plant differences in physical mortality should be analyzed.* Differences in mortality of entrained species are also related to power plant design and operational characteristics. Little is known about specific differences among plants which cause such variation in effects on the biota. Field investigations using standardized methods are needed to accurately compare organism survivals and to relate these to variations in physical stresses imposed during entrainment.

2. *Entrainment sampling methodology should be standardized.* Quantitative samples are necessary both at the intake and discharge in order to estimate survival rates among entrained organisms. If new gear is introduced, data should be included on its efficiency relative to a standard gear. The total number of organisms per cubic meter of water at the intake should be statistically comparable to that at the discharge side of condensers. Sampling net or pump-caused mortalities must be kept to a minimum and must be comparable in both the intake and discharge samples. Mortalities induced by collection devices should be factored into the mortality assessments, following procedures such as those provided by O'Connor and Schaffer (in press).

3. *An entrainment sampling program should have a defensible statistical design which establishes confidence limits, sampling error, variance, power curves, etc.* The adequacy of sampling only a given fraction of the total intake flow per unit time should be statistically demonstrated. Complete physical measurements should be taken during sampling. The developmental stages and the sizes of organisms should be recorded during all determinations of the components of inner-plant mortality in order to estimate relative vulnerability.

Day and, especially, night studies should be conducted to determine if the rates of entrainment of organisms and size-related mortality vary during a 24 hr period. Sampling should be carried out over the entire spawning season of the potentially vulnerable species, at a suitable interval to detect significant temporal and spatial changes in species composition and relative abundance.

Modeling should be used to predict the entrainment impact of proposed new power plants or modifications of existing plants. The potential population of adults lost based on losses of entrained meroplankton should be estimated. New study designs are needed to assess the impact of effluent plume entrainment, especially that due to recent high-speed jet diffuser designs.

C. Operation

1. *Coordinating plant activities with organism densities is one approach which can help to alleviate the entrainment problem.* For example, planning plant shutdowns for refueling or maintenance to coincide with high peaks of egg and larval density; varying pumping rates, day vs. night or flood vs. ebb tides, to correspond with known densities; and, possibly, using multiple intakes in different areas or depths so as to draw water from areas of low organism densities may reduce the adverse effects of entrainment.
2. *The impact of power plants can be minimized by siting in nonproductive areas and, in many cases, utilizing closed cooling systems, i.e., lower intake volumes, since the significance of the mortality of entrained biota is directly related to the volume entering the intake.* This low volume concept is presently the only immediately effective approach to minimizing the adverse effects on entrained organisms if high mortality is assumed or observed during plant passage.

Also, it may be possible to increase condenser ΔT's in conjunction with the lowering of intake volumes, especially when the physical damage component of the mortality dominates as appears to be the case with entrained meroplankton and juvenile fishes.

3. *It may be possible to maintain high flow rates (and thereby low ΔT's) by redesigning pumps to decrease internal physical stress.* However, this is a long-range alternative requiring additional knowledge of the physical tolerances of entrained organisms as well as improved ability to assess physical stress levels within various pump designs.

ACKNOWLEDGMENTS

Appreciation is extended to Luise Davis, Richard Nugent, Alice Lawson, Vera Percy, Terry Rojahn and the Ecological Sciences Division of NUS Corporation for their helpful suggestions on the manuscript and for providing editorial and typing assistance. This chapter is also identified as part of Contribution No. 737, Chesapeake Biological Laboratory, Solomons, Maryland.

REFERENCES

Adams, J. R. 1968. Thermal effects and other considerations at steam electric plants. A survey of studies in the marine environment. Pacific Gas and Electric Co., Dept. of Eng. Res. Rept. No. 6934. Cited in Schubel (1975).

AEC. 1972. Proposed issuance of an operating license to the Boston Edison Company for the Pilgrim Nuclear Power Station. Docket No. 50-293. Draft environmental impact statement. Cited in Ulanowicz (1975).

AEC. 1973. Draft environmental statement related to the Oyster Creek nuclear generating station. Docket No. 50-219.

AEC. 1974. Final environmental statement. Millstone Nuclear Power Station - Unit 3. Docket No. 50-423.

Alden, R. W., III, F. J. S. Maturo, Jr. and W. Ingram III. 1976. Interactive effects of temperature, salinity and other factors on coastal copepods. *In* Thermal Ecology II. Proc. of Symp., Savannah River Ecology Lab., Inst. of Ecology, Univ. Georgia. April 2-5, 1975. ERDA Sym. Ser. 40:336-348.

American Nuclear Society. 1974. Entrainment: guide to steam electric power plant cooling system siting, design, and operation for controlling damage to aquatic organisms. ANS - 18.3 Committee. Draft No. 8.

Arnold, D. E., and Pennsylvania Cooperative Fishery Research Unit. 1975. Biological perspectives on steam-electric generating station discharges to the Great Lakes. Contract No. AT(11-1)-2141, EPA, Great Lakes Res. Div., Univ. of Michigan, Ann Arbor, Mich.

Austin, H. M., J. Dickinson, and C. Hickey. 1973. An ecological study of ichthyoplankton at the Northport power station, Long Island, New York. New York Ocean Sci. Lab., Montauk, N.Y. Rept. to Long Island Lighting Co. Cited in Beck and Lackie (1974).

Battle, H. I. 1944. Effects of dropping on the subsequent hatching of teleostean ova. J. Fish. Res. Board Can. 6:252-256.

Beck, A. D., and N. F. Lackie. 1974. Effects of passage of marine animals through power plant cooling water system. Presented at Am. Fish. Soc. Annual Meeting, Hawaii.

Beck, A. D., and D. C. Miller. 1974. Analysis of inner plant passage of estuarine biota. Presented at the Am. Soc. Civil Eng. Power Div. Spec. Conf., Boulder, Colo., Aug. 12-14.

Beck, A. P., G. V. Poje, and W. T. Waller. 1975. A laboratory study on the effects of exposure of some entrainable Hudson River biota to hydrostatic pressure regimens calculated for the proposed Cornwall pump-storage plant, p. 167-204. *In*

S. B. Saila (ed.), Fisheries and energy production, a symposium. Lexington Books, Lexington, Mass.

Benda, R. S. 1972. Thermal effects studies at the Palisades Nuclear Plant. Preliminary findings from the first few months of operation. Presented at Am. Fish. Soc. Nat. Meetings, Hot Springs, Ark. Cited in Davies and Jensen (1974).

Bongers, L. H., A. J. Lippson, and T. T. Polgar. 1973. An interpretive summary of the 1973 Morgantown entrainment studies. Rept. to Potomac Electric Power Co. by Martin Marietta Labs., Baltimore, Md.

Briand, F. J-P. 1975. Effects of power plant cooling systems on marine phytoplankton. Mar. Biol. 33:135-146.

Carlson, F. T., and J. A. McCann. 1969. Hudson River fisheries investigations (1965-1968). Rept. to Consolidated Edison Co. by the Hudson River Policy Comm.

Carpenter, E. J., S. J. Anderson, and B. B. Peck. 1972. Entrainment of marine plankton through Millstone Unit 1. 3rd and 4th semi-annual rept., Woods Hole Oceano. Instit., Woods Hole, Mass.

Carpenter, E. J., B. B. Peck, and S. J. Anderson. 1974. Survival of copepods passing through a nuclear power station on northeastern Long Island Sound, U.S.A. Mar. Biol. 24:49-55.

Clark, J., and W. Brownell. 1973. Electric power plants in the coastal zone: environmental issues. Spec. Publ. No. 7, Am. Litt. Soc., Highlands, N.J.

Consumers Power Company. 1972a. Palisades Plant. Docket No. 50-255. Special report No. 4, Environmental impact of plant operation up to July 1972. Consumers Power Co., Jackson, Mich. Cited in Davies and Jensen (1974).

Consumers Power Company. 1972b. Palisades Plant. Docket No. 50-255. Semi-annual operations report No. 4, 1 July-Dec. 1972. Consumers Power Co., Jackson, Mich. Cited in Davies and Jensen (1974).

Consumers Power Company. 1973. Palisades Plant. Semi-annual operations report No. 5. January 1-June 30, 1973.

Copeland, B. J., R. G. Hodson, and W. S. Birkhead. 1975. Entrainment and entrainment mortality at the Brunswick Nuclear Power Plant. Part I of: Report on entrainment and entrainment mortality of zooplankton and larvae, and impingement and movement of fish. Rept. to Carolina Power and Light, Raleigh, N.C.

Coutant, C. C. 1970. Biological aspects of thermal pollution. I. Entrainment and discharge canal effects. C.R.C. Crit. Rev. Envir. Control 1:341-381.

Coutant, C. C., and R. J. Kedl. 1975. Survival of larval striped bass exposed to fluid-induced and thermal stresses in a stimulated condenser tube. Publ. No. 637, Oak Ridge Natl. Lab., Oak Ridge, Tenn. (Not a final report).

Davies, R. M., and L. D. Jensen. 1974. Effects of entrainment of zooplankton at three Mid-Atlantic power plants. Rept. No. 10 prepared for Elec. Power Res. Inst. cooling water discharge research project RP-49. EPRI Publ. No. 74-049-00-1NUS.

Davies, R. M., C. H. Hanson and L. D. Jensen. 1976. Entrainment of estuarine zooplankton into a mid-Atlantic power plant: delayed effects. *In*: Thermal Ecology II. Proc. Symp., Savannah River Ecology Lab., Instit. of Ecology, Univ. Georgia. April 2-5, 1975. ERDA Symp. Ser. 40: 349-352.

Davis, H. S. 1953. Culture and diseases of game fishes. Univ. of Calif. Press. Berkeley, Calif.

Edsall, T. A., and T. G. Yocum. 1972. Review of recent technical information concerning the adverse effects of once-through cooling on Lake Michigan. Prepared for the Lake Michigan Enforc. Conf., Sept. 19-21, 1972, Chicago. U.S. Fish Wildl. Serv., Great Lakes Fish Lab., Ann Arbor, Mich.

Edsall, T. A. 1976. Electric power generation and its influence on Great Lakes fish, p. 453-462. *In* Proc. of the Second Conf. on the Great Lakes. Great Lakes Comm.

Emadi, H. 1973. Yolk-sac malformation in Pacific salmon in relation to substrate, temperature, and water velocity. J. Fish. Res. Board Can. 30:1249-1250.

EPA. 1972a. Proceedings in the matter of pollution of Mount Hope Bay and its tributaries.

EPA. 1972b. Lake Michigan entrainment studies, Big Rock Nuclear Power Plant, Escanaba Power Plant, November-December 1971. Grosse Ile Laboratory Working Rept. No. 1. Mimeo. Cited in Arnold and Pa. Coop. Fish. Res. Unit (1975).

EPA. 1973. Reviewing environmental impact statements - power plant cooling systems, engineering aspects. EPA 660/2-73-016. Natl. Envir. Res. Ctr., Corvallis, Ore.

Eraslen, A. H., W. Van Winkle, R. D. Sharp, S. W. Cristensen, C. P. Goodyear, R. M. Rush, and W. Fulkerson. 1976. ORNL/ NUREG-8 SPEC. Env. Sci. Div. Publ. No. 766, Oak Ridge Natl. Lab., Oak Ridge, Tenn.

Flemer, D. A., C. W. Keefe, L. Hicks, D. R. Heinle, M. C. Grote, R. P. Morgan, W. Gordon, L. Dorsey, and J. A. Mihursky. 1971a. The effects of steam electric station operation on entrained organisms, p. 1-15. *In* J. A. Mihursky and A. J. McErlean (co-principal investigators), Postoperative assessment of the effects of estuarine power plants. Nat. Res. Instit., Ref. No. 71-24b.

Flemer, D. A., D. R. Heinle, R. P. Morgan, C. W. Keefe, M. C. Grote, and J. A. Mihursky. 1971b. Preliminary report on the effects of steam electric station operations on entrained organisms. A progress report to the Maryland Dept. Nat. Res. under the post-operative assessment of effects of estuarine power plants grant. Univ. of Maryland, College Park, Md.

Grosse Ile Laboratory. 1972. Lake Michigan entrainment studies, Big Rock nuclear power plant, Escabana power plant, Nov.-Dec. 1971. Rept. No. 1, EPA Ofc. Res. and Monitor., Grosse Ile, Mich.

Gurtz, M. E., and C. M. Weiss. 1972. Field investigations of the response of phytoplankton to thermal stress. Envir. Study Program, Duke Power Co., ESE Publ. No. 321, Univ. of N. Carolina, Chapel Hill, N.C.

Hardy, C. D. 1971. Dissolved oxygen survey at Northport, p. 116-119. *In* G. C. Williams, Studies on the effects of a steam-electric generating plant on the marine environment Northport, New York. Tech. Rept. No. 9, Marine Sci. Res. Ctr., State Univ. New York, Stony Brook, N.Y.

Hayes, F. R. 1949. The growth, general chemistry and temperature relations of salmonid eggs. Q. Rev. Biol. 24:281.

Heinle, D. R. 1976. Effects of passage through power plant cooling systems. Envir. Pollut. Vol. II, pp. 39-57.

Hess, K. E., M. P. Sissenwine, and S. B. Saila. 1975. Simulating the impact of the entrainment of winter flounder larvae, p. 1-29. *In* S. B. Saila (ed.), Fisheries and energy production, a symposium. Lexington Books, Lexington, Mass.

Icanberry, J., and J. Adams. In press. Zooplankton survival in cooling water systems of four thermal power plants for the California coast - interim report - March 1971-January 1972. *In* Davies and Jensen (1974).

Icanberry, J. W. 1973. Zooplankton survival in cooling water systems of four California coastal power plants, August 1971-July 1972. Presented at 36th annual meeting ASLO, June 10-14, Univ. Utah, Salt Lake City.

Industrial Bio-Test Laboratories, Inc. 1971. Phytoplankton study, preliminary report, March-July, 1970. IBT No. W8956, Project III. Cited in Argonne National Laboratory (1972).

Industrial Bio-Test Laboratories, Inc. 1972. Intake-discharge experiments at Waukegan generating station. Rep. to Commonwealth Edison Co., Proj. XI, IBT No. W9861, Biol. Sec.

Jensen, L. D. 1977. Effects of thermal discharges upon aquatic organisms in estuarine waters with discussion of limiting factors. *In* Estuarine Pollution Control and Assessment Vol.

1. USEPA Office of Water Planning and Standards, Washington, D.C. pp. 359-372.

Kinne, O. (ed.). 1972. Marine ecology. Vol. 1, Part 3. John Wiley and Sons, Ltd., N.Y. Cited in Beck et al. (1975).

Knight-Jones, E. W., and E. Morgan. 1966. Responses of marine animals to changes in hydrostatic pressure. Oceanog. Marine Biol. Ann. Rev. 4:267-299. Cited in Beck et al. (1975).

Knutson, K. M., S. R. Berguson, D. L. Rastetter, M. W. Mischuk, F. B. May, and G. M. Kuhl. 1976. (manuscript). Seasonal pumped entrainment of fish at the Monticello, Mn. Nuclear Power Installation. Dept. of Biol. Sci., St. Cloud State Univ., St. Cloud, Minn.

Lackey, J., and E. Lackey. In press. Thermal effects at Turkey Point-a study. *In* L. D. Jensen (ed.). Proceedings of the entrainment and intake screening workshop, Feb. 1973. Johns Hopkins Univ. Cooling Water Res. Proj., Johns Hopkins Univ., Baltimore. Cited in Davies and Jensen (1974).

Lauer, G. J., W. T. Waller, D. W. Bath, W. Meeks, R. Heffner, T. Ginn, L. Zubarik, P. Bibko, and P. C. Storm. 1973. Entrainment studies on Hudson River organisms. New York Univ. Med. Center, Lab. for Environ. Studies Rept. Nov. 1973: 208. Cited in Beck and Lackie (1974).

Lauer, G. J., W. T. Waller, D. W. Bath, W. Meeks, R. Heffner, T. Ginn, L. Zubarik, P. Bibko, and P. C. Storm. 1974. Entrainment studies on Hudson River organisms, p. 37-82. *In* L. D. Jensen, 1974. Proceedings of the second entrainment and intake screening workshop. Rept. No. 15, The Johns Hopkins Univ., Baltimore, Md.

Leitritz, E. 1963. Trout and salmon culture. Calif. Dept. Fish Game Fish. Bull. 107.

Marcy, B. C., Jr. 1971. Survival of young fish in the discharge canal of a nuclear power plant. J. Fish. Res. Board Can. 28:1057-1060.

Marcy, B. C., Jr. 1973. Vulnerability and survival of young Connecticut River fish entrained at a nuclear power plant. J. Fish. Res. Board Can. 30(8):1195-1203.

Marcy, B. C., Jr. 1975. Entrainment of organisms at power plants with emphasis on fishes-an overview, p. 89-106. *In* S. B. Saila (ed.), Fisheries and energy production, a symposium. Lexington Books, Lexington, Mass.

Marcy, B. C., Jr. 1976. Planktonic fish eggs and larvae of the lower Connecticut River and the effects of the Connecticut Yankee plant including entrainment, p. 115-139. *In* D. Merriman and L. Thorpe (eds.), The Connecticut river ecological study: the impact of a nuclear power plant. Am. Fish. Soc. Monogr. No. 1.

Markowski, S. 1962. Faunistic and ecological investigations in Cavendish Dock, Barrow-In-Furness. J. Anim. Ecol. 31:46-51. Cited in Beck and Lackie (1974).

Massengill, R. R. 1976. Entrainment of zooplankton at the Connecticut Yankee Plant, p. 55-59. *In* D. Merriman and L. Thorpe (eds.), The Connecticut River ecological study. The Am. Fish. Soc. Monogr. No. 1.

Mathur, D. S., and G. M. Yazdani. 1969. Observations on a deformed specimen of *Heteropneustes fossilis* Siluriformes Heteropneustidae. Sci. Cult. 35(9):490-491.

Matlak, O. 1970. Vorlaufinger Bericht uber Missbildung des Kopfes beim Karpfenstrich. Acta Hydrobiol. 12(4):391-398.

Maturo, F. J., R. Alden, and W. Ingram. 1974. Effects of power plant entrainment on major species of copepods, p. 69-101. *In* Crystal River power plant environmental considerations. Final report to the Interagency Research Advisory Committee, Vol. IV. Florida Power Co., St. Petersburg, Fla. Cited in Weiss (unpublished ms).

McNaught, D. C. 1972. The potential effects of condenser passage on the entrained zooplankton at Zion Station. Prepared statement at the Lake Michigan Enforcement Conf., Sept. 21, 1972, Chicago, Ill.

Middlebrook, K. 1975. Zooplankton entrainment, p. 2.3.2-1 through 2.3.2-16. *In* Environmental monitoring and ecological studies program for the Prairie Island Nuclear Generating Plant. NSP 1974 annual report. Northern States Power Co., Minneapolis, Minn.

Mihursky, J. A., and L. C. Dorsey. 1973. Macroplankton, p. 152-170. *In* The effects of Morgantown Steam Electric Station operations on organisms pumped through the cooling water system. Final rept. to Maryland Dept. of Nat. Res., Feb. 1973. Cited in Beck and Lackie (1974).

Morgan, R. P., and R. G. Stross. 1969. Destruction of phytoplankton in the cooling water supply of a steam electric station. Ches. Sci. 10(3-4):165-171.

Morgan, R. P., R. E. Ulanowicz, V. J. Rassin, L. A. Noe, and G. B. Gray 1973. Effects of water movement on eggs and larvae of striped bass and white perch. Nat. Res. Instit., Chesapeake Biol. Lab., Univ. of Maryland, NRI Ref. No. 73-111.

Morgan, R. P., R. E. Ulanowicz, V. J. Rasin, L. A. Noe, and G. B. Gray. 1976. Effects of shear on eggs and larvae of the striped bass, *Morone saxatilis,* and the white perch, *M. americana*. Trans. Am. Fish. Soc. 105:149-154.

NALCO Environmental Sciences. 1976. Report to Northern Indiana Public Service Company - Dean H. Mitchell Station. Section 316(a) Demonstration. Hammond, Ind.

Nawrocki, S. S. 1977. A study of fish abundance in Niantic Bay with particular reference to the Millstone Point Nuclear Power Plant. M.S. Thesis. Univ. of Conn., Storrs, Conn.

New York University Medical Center. 1975. The effects of changes in hydrostatic pressure on some Hudson River biota. Instit. of Envir. Medicine. Prog. Rep. for 1975. Prepared for Consolidated Edison Company of New York, Inc., New York, N.Y.

NRC. 1976. Final environmental statement related to the selection of the preferred closed cycle cooling system at Indian Point Unit No. 2. Docket No. 50-247. NUREG-0042.

O'Connor, J. M. and S.A. Schaffer. 1977. The effects of sampling gear on the survival of striped bass ichthyoplankton. Ches. Sci. (In press).

Oglesby, R. T., and D. J. Allee. 1969. Ecology of Cayuga Lake and the proposed Bell station (nuclear powered). Water Res. Mar. Sci. Cent., Cornell Univ., Ithaca, N.Y., No. 27:315-328.

Post, G., D. V. Power, and T. M. Kloppel. 1973. Effects of physical shock on developing salmonid eggs. Trans. Am. Nucl. Soc. 17:27-28.

Prager, J. C., E. W. Davey, J. H. Gentile, W. Gonzlalez, R. L. Steele, M. D. Lair, C. I. Weber, and S. L. Bargo. 1970. A study of Biscayne Bay plankton affected by the Turkey Point thermal electric generating plant during July and August, 1970. EPA Natl. Mar. Water Qual. Lab. Spec. Rept., November. W. Kingston, R.I.

Prager, J. C., R. L. Steele, W. Gonzlalez, R. Johnson, and R. L. Highland. 1971. Survey of benthic microbiota and zooplankton conditions near Florida Power and Light Company's Turkey Point Plant, August 23-27, 1971. Natl. Mar. Water Qual. Lab., W. Kingston, R.I.

Profitt, M. A. 1969. Effects of heated discharge upon aquatic resources of the White River at Petersburg, Indiana. Indiana Univ. Water Res. Ctr., Rep. of Invest. No. 3.

Restaino, A. L., D. G. Redmond and R. G. Otto. 1975. Entrainment study at Zion Station. *In* Compilation of special reports on the effects of Zion Station operation on the biota in southwestern Lake Michigan. 1975. Report to Commonwealth Edison Company, Chicago, Illinois by Industrial Bio-Test Lab., Inc.

Rollefsen, G. 1930. Observation on cod eggs. Rapp. Cons. Explor. Mer. 65:31-34.

Sandine, P. H. 1973. Zooplankton of Barnegat Bay: the effects of the Oyster Creek Nuclear Power Plant. M.S. Thesis. Rutgers Univ., New Brunswick, N.J. Cited in Beck and Lackie (1974).

Schubel, J. R. 1973. Effects of exposure to time-excess temperature histories typically experienced at power plants on the hatching success of fish eggs. Spec. Rep. No. 32-PPRP-4, Ref. No. 73.11. Chesapeake Bay Instit., The Johns Hopkins Univ., Baltimore, Md.

Serchuk, F. M. 1976. The effects of the Ludington Pumped Storage Power Project on fish passage through pump-turbines and on fish behavior patterns. Ph.D. Disser. Mich. State Univ., Ann Arbor, Mich.

Sleigh, M. A., and A. G. McDonald (eds.). 1972. The effects of pressure on organisms. Symposium of the Society of Experimental Biology. Academic Press, N.Y. Cited in Beck et al. (1975).

Smith, R. A., and A. S. Brooks. 1971. Primary production, p. 33-37. *In* R. W. Koss (ed.), An interim report on environmental responses to thermal discharges from Marshall steam station, Lake Norman, North Carolina. Cooling Water Discharge Proj. RP-49, Edison Elec. Instit., The Johns Hopkins Univ., Baltimore, Md.

Smith, R. A., and L. D. Jensen. 1974. Effects of condenser destruction of algae on dissolved oxygen levels in the James River, p. 123-129. *In* L. D. Jensen. Proceedings of the second entrainment and intake screening workshop. Rept. No. 15, The Johns Hopkins Univ., Baltimore, Md.

Tarzwell, C. M. 1972. An argument for open ocean siting of coastal thermal electric plants. J. Environ. Qual. 1(1):89-91.

Teleki, G. C. 1976. The incidence and effect of once-through cooling of young-of-the-year fishes at Long Point Bay, Lake Erie: a preliminary assessment. *In* Thermal Ecology II Proc.

of Symposium, Savannah River Ecology Lab. Instit. Ecology. Univ. of Georgia. April 2-5, 1975. ERDA Sym. Sec. 40:387-393.

Tsvetkov, V. I., D. S. Pavlov, and V. K. Nezdoliy. 1972. Changes of hydrostatic pressure lethal to young of some freshwater fish. J. Ichthyol. 12:307-308. Cited in Beck et al. (1975).

Ulanowicz, R. 1975. The mechanical effects of water flow on fish eggs and larvae, p. 77-87. *In* S. B. Saila (ed.). Fisheries and energy production, a symposium. Lexington Books, Lexington, Mass.

University of Wisconsin-Milwaukee. 1972. Environmental studies at the Point Beach NUclear Power plant. April 1972. Department of Botany. Rep. No. PBR3.

Voigtlander, C. W. 1974. Toward the assessment of impacts of entrainment: estimation of populations and entrainment on a pre-volume basis. Fish. and Waterfowl Res. Branch Tenn. Valley Auth. (manuscript).

Williams, G. C. 1971. Studies on the effects of a steam-electric generating plant on the marine environment at Northport, New York. Tech. Rep. No. 9, Mar. Sci. Res. Ctr. State Univ. New York, Stony Brook, N.Y.

Wisconsin Public Service Corp. 1974. Preoperational thermal monitoring program of Lake Michigan near the Kewaunee Nuclear Power plant. Jan.-Dec. 1973. Third Annual Rep., Green Bay, Wisc. March 1974.

Wolke, R. E., G. R. Boucks, and R. K. Stroud. 1975. Gas-bubble disease: a review in relation to modern energy production, p. 239-265. *In* S. B. Saila, Fisheries and energy production, a syposium. Lexington Books. Lexington, Mass.

Zimmerman, A. M. (ed.). 1970. High pressure effects on cellular processes. Academic Press, N.Y. Cited in Beck et al. (1975).

CHAPTER 5. CUMULATIVE EFFECTS
A FIELD ASSESSMENT

ALLAN D. BECK
and
THE COMMITTEE ON ENTRAINMENT

I. INTRODUCTION

Observations at operating power plants show that many organisms that pass through their cooling systems do not survive. The effects of the physical, chemical, and thermal stresses act in concert on the entrained biota. Because the combinations of site, biota, receiving water characteristics and plant operating conditions are almost limitless, the effects are plant specific. At a plant the dominant source of damage may change seasonally and with variations in the mode of plant operation. Thermal stresses may control mortality of entrained organisms during periods of high ambient temperature, while chemical stresses may dominate during periods of heavy chlorination. In many cases physical stresses including pressure changes, shear, turbulence, impact, and abrasion are the principal causes for mortality of entrained organisms.

II. FIELD OBSERVATIONS

Results of field studies at 14 power plants with once-through cooling systems are summarized in Tables 1, 1a. Data from different plants must be compared with prudence. Sampling methods, incubation techniques, duration of observational periods, and even the criterion for assessing death vary among investigators. In some studies data are available for only a few individuals, and even when data are abundant they frequently have not been subjected to rigorous statistical analysis. For details on sampling and analysis, the reader is referred to the original papers.

Tables 1, 1a are intended to provide insight into the relative importance of the several kinds of stresses experienced during pump entrainment in producing mortality of entrained organisms. Despite the limitations of these data, certain conclusions can be drawn.

In those cases where mortality of entrained ichthyoplankton and juvenile fishes has been apportioned among the several kinds of stresses--thermal, chemical, and physical--physical stresses usually dominate. In every study of entrained ichthyoplankton and juvenile fishes where the causes of mortalities have been identified, physical stresses had a greater effect on mortality than did temperature. In all cases but one, physical stresses were more important than chemical stresses in producing mortality. According to Lauer et al. (1973) the total mortality fraction of white perch and striped bass larvae entrained by the Indian Point plant on the Hudson River during periods of chlorination was about 50%, with 60% of this total attributed to biocidal stresses (chlorination), 30% to physical stresses, and the remainder--10%--to thermal stresses. Of the other ichthyoplankton studies summarized in Table 1, from about 80 to 100% of the total observed mortalities--mortalities that ranged from about 0 to 100% of the organisms entrained--were attributed to physical stresses.

According to the studies summarized in Tables 1, la the relative importance of chlorination in causing mortality of zooplankton may be greater than for larval and juvenile fishes. The effects of chlorination on mortality were greater than either the thermal or physical stresses in more than 75% of those studies in which it is possible to apportion the mortality among the various classes of stress. Physical stresses usually had a greater impact on mortality than did temperature. Carpenter et al. (1974) reported, for example, that the mortality of copepods at the Millstone Point (Conn.) power plant on eastern Long Island Sound ranged from 67-83%. Carpenter (personal communication) estimated that 70% of the observed mortality was caused by physical stresses.

In summary, although a relatively large number of field studies have been conducted to determine the mortality of entrained organisms few of the investigators have apportioned the total mortality among the several classes of stresses. For those that have, physical stresses dominate most frequently in causing mortality of ichthyoplankton and juvenile fishes. The data are sparse and these generalizations tenuous. Entrainment studies should attempt to apportion the total mortalities among the physical, thermal, and chemical stresses. Without this information, the data are of little value in selecting plant operating criteria to minimize the total number of organisms killed by entrainment (see Chapter 6).

Table 1 Effects of Plant Passage (Entrainment) on Marine/Estuarine Fishes
Mortalities Indicated are in Excess of Mortalities Occurring in Control Samples

SPECIES/ORGANISMS & LIFE STAGE	GEOGRAPHIC LOCATION & PLANT DESIGNATION	EFFECTS % Mortality of Entrained Organisms					REFERENCES & REMARKS
		Physical	Thermal	Chemical	Other	Total	
Alosa aestivalis (Blueback herring) & Alosa pseudoharengus (Alewife) larvae & early juvenile	Connecticut Yankee Connecticut River Haddam Neck, CT	≈57.0(ab)	≈14.0(ab)	(c)	-	≈72.0(bd)	Marcy (1973) a. Calculated from original author's estimates of relative effects. b. Discharge canal temp. 28 to 29°C c. No apparent additional effect. d. Thermal exposure approx. 93 sec in condenser and 50-100 min in discharge canal. NOTE: Mortality percentages are for a combined group of Alosa aestivalis and Alosa pseudoharengus. It is assumed mortality for both species is approximately similar.
Ammodytes dubius (Northern sand lance) larvae	Millstone Point Long Island Sound Waterford, CT	10.4 to 23.9	-	-	-	10.4(ab) to 23.9	Nawrocki (1977) a. Total % damaged, (author). b. Varies seasonally.
Anchoa mitchilli (Bay anchovy) larvae	Millstone Point Long Island Sound Waterford, CT	8.5 to 23.1	-	-	-	8.5(ab) to 23.1	Nawrocki (1977) a. Total % damaged, (author). b. Varies seasonally.

Table 1 (Continued)

Anguilla rostrata *(American eel)* *Larvae*	*Millstone Point* *Long Is. Sound* *Waterford, CT*	*0*	-	-	-	*0(a)*	*Nawrocki (1977)* *a. Total % damaged, (author).*
Anguilla rostrata *(American eel)* *juvenile*	*Indian Point (a)* *Hudson River, NY*	*0*	*0*	*0*	*0*	*0*	*Lauer et al. (1973)* *a. Salinity at site ranges from 0 °/∘∘ to 7 °/∘∘.*
Brevoortia tyrannus *(Atlantic menhaden)* *larvae*	*Long Island Sound,* *Northport, NY*	*27.6*	-	-	-	*>27.6*	*Austin, Dickinson & Hickey (1973)* *Mortality expressed as % macerated.*
Cynoscion regalis *(Weakfish)* *larvae*	*Long Island Sound,* *Northport, NY*	*33.3(a)*	-	-	-	*>33.3*	*Austin, Dickinson & Hickey (1973)* *a. Data based on <100 individuals. Mortality expressed as % "macerated".*
Enchelyopus cimbrius *(Fourbeard rockling)* *larvae*	*Millstone Point* *Long Island Sound* *Waterford, CT*	*18.5* *to* *23.8*	-	-	-	*18.5(ab)* *to* *23.8*	*Nawrocki (1977)* *a. Total % damaged, (author).* *b. Varies seasonally.*
Lumpenus lumpretaeformis *(Snakeblenny)* *larvae*	*Millstone Point* *Long Island Sound* *Waterford, CT*	*15.0* *to* *23.3*	-	-	-	*15.0(ab)* *to* *23.3*	*Nawrocki (1977)* *a. Total % damaged, (author).* *b. Varies seasonally.*
Menidia menidia *(Atlantic silverside)* *larvae*	*Millstone Point* *Long Island Sound* *Waterford, CT*	*42.3* *to* *46.5*	-	-	-	*42.3(ab)* *tQ* *46.5*	*Nawrocki (1977)* *a. Total % damaged, (author).* *b. Varies seasonally.*

Table 1 (Continued)

SPECIES/ORGANISMS & LIFE STAGE	GEOGRAPHIC LOCATION & PLANT DESIGNATION	EFFECTS % Mortality of Entrained Organisms					REFERENCES & REMARKS
		Physical	Thermal	Chemical	Other	Total	
Morone americana (White perch) larvae & early juvenile	Connecticut Yankee Connecticut River Haddam Neck, CT.	80(a)	20(ab)	(c)	-	100(bd)	Marcy (1973) a. Calculated from original author's estimate of relative mortalities. b. Discharge canal temp. ranged from 28 to 33.5°C. Exposure time approx. 93 sec in condenser and 50-100 min in discharge canal. c. No apparent additional effects. d. All data based on <100 individuals.
Morone spp. (a) (White perch & Striped bass) larvae	Indian Point (d) Hudson River, NY	23.0(b)	4.0(b)	33.0(b)	21.0(bc)	54.0(b)	Lauer et al. (1973) a. "Pooled" sample with Morone saxatilis and Morone americana. b. Includes dead & stunned. c. Combined effects: mechanical and thermal. d. Salinity at site ranges from 0 °/oo to 7 °/oo.
Myoxocephalus aenaeus (grubby) larvae	Millstone Point Long Island Sound Waterford, CT	7.5 to 15.6	-	-	-	7.5(ab) to 15.6	Nawrocki (1977) a. Total % damaged, (author). b. Varies seasonally.
Peprilus triacanthus (Butterfish) larvae	Millstone Point Long Island Sound Waterford, CT	41.1	-	-	-	41.1(a)	Nawrocki (1977) a. Total % damaged, (author).

Table 1 (Continued)

Prionotus carolinus *(Northern searobin)* *larvae*	*Millstone Point Long Island Sound Waterford, CT*	*27.7*	-	-	-	*27.7(a)*	*Nawrocki (1977)* *a. Total % damaged, (author).*
Pseudopleuronectes americanus *(Winter flounder)* *larvae*	*Millstone Point Long Island Sound Waterford, CT*	*8.7 to 14.3*	-	-	-	*8.7(ab) to 14.3*	*Nawrocki (1977)* *a. Total % damaged, (author).* *b. Varies seasonally.*
Scomber scombrus *(Atlantic mackerel)* *larvae*	*Millstona Point Long Island Sound Waterford, CT*	*22.6 to 31.4*	-	-	-	*22.6(ab) to 31.4*	*Nawrocki (1977)* *a. Total % damaged, (author).* *b. Varies seasonally.*
Scophthalmus aquosus *(Windowpane flounder)* *larvae*	*Millstone Point, Long Island Sound Waterford, CT*	*16.8 to 20.5*	-	-	-	*16.8(ab) to 20.5*	*Nawrocki (1977)* *a. Total % damaged, (author).* *b. Varies seasonally.*
Stenotomus chrysops *(Scup)* *larvae*	*Millstone Point Long Island Sound Waterford, CT*	*14.3 to 37.8*	-	-	-	*14.3(ab) to 37.8*	*Nawrocki (1977)* *a. Total % damaged, (author).* *b. Varies seasonally.*
Syngnathus fuscus *(Northern pipefish)* *larvae*	*Millstone Point Long Island Sound Waterford, CT*	*0*	-	-	-	*0(a)*	*Nawrocki (1977)* *a. Total % damaged, (author).*
Tautoga onitis *(Tautog)* *larvae*	*Long Island Sound, Northport, NY*	*28.6(a)*	-	-	-	*>28.6(a)*	*Austin, Dickinson & Hickey (L973)* *a. Data based on <100 individuals. Mortality expressed as % "macerated".*
Tautoga onitis *(Tautog)* *larvae*	*Millstone Point Long Island Sound Waterford, CT*	*36.8 to 42.0*	-	-	-	*36.8(ab) to 42.0*	*Nawrocki (1977)* *a. Total % damaged, (author).* *b. Varies seasonally.*

Table 1 (Continued)

SPECIES/ORGANISMS & LIFE STAGE	GEOGRAPHIC LOCATION & PLANT DESIGNATION	EFFECTS % Mortality of Entrained Organisms					REFERENCES & REMARKS
		Physical	Thermal	Chemical	Other	Total	
Tautogolabrus adspersus (Cunner) larvae	Millstone Point Long Island Sound Waterford, CT	25.6 to 50.8	-	-	-	25.6(ab) to 50.8	Nawrocki (1977) a. Total % damaged, (author). b. Varies seasonally.
Tautogolabrus adspersus (Cunner) larvae	Long Island Sound Northport, NY	57.1(a)	-	-	-	>57.1(a)	Austin, Dickinson & Hickey (1973) a. Data based on <100 individuals. Mortality expressed as % "macerated".
Clupeids (spp) larvae	Millstone Point Long Island Sound Waterford, CT	62.5	-	-	-	62.5(a)	Nawrocki (1977) a. Total % damaged, (author).
Fishes (cold, temperate, Atlantic coast, summer species) June and July larvae & early juvenile	Connecticut Yankee Connecticut River, Haddam Neck, CT	70.5 to 86.8	14.8(ab) to 20.0	5.2 to 10.3	-	74.1(b) to 100.0	Marcy (1973) a. Calculated from original author's estimates of relative effect. b. Discharge canal temp. ranged from 28 to 35°C. Exposure time approximately 93 sec in condenser and 50-100 min in discharge canal.

Table 1a Effects of Plant Passage (Entrainment) on Marine/Estuarine Zooplankton and Phytoplankton
Mortalities Indicated are in Excess of Mortalities Occurring in Control Samples

SPECIES/ORGANISMS & LIFE STAGE	GEOGRAPHIC LOCATION & PLANT DESIGNATION	EFFECTS % Mortality of Entrained Organisms					REFERENCES & REMARKS
		Physical	Thermal	Chemical	Other	Total	
Acartia clausi (Copepod) adult	Millstone Point, Long Island Sound Waterford, CT	-	-	-	91.0 (a)	92.0 to 99.0	U.S. Atomic Energy Commission Directorate of Licensing- Docket No. 50-423, February 1974. a. Combined effects: mechanical and thermal.
Acartia spinata (Copepod) adult	Turkey Point, Biscayne Bay, FL	-	-	-	-	67.3(a) to 100.0	Prager et al. (1970) a. Passage time intake through pumps and condenser to discharge approx. 1-2 min.
Acartia tonsa (Copepod) adult	Potomac River, Morgantown, MD	40	10	-	-	50(a)	Lackie (1973) Unpublished data. a. Other zooplankton species also included in sample. Acartia tonsa was the large majority, approx. 70-90%.
Acartia tonsa (Copepod) adult	Indian Point (c) Hudson River, NY	-	-	12.3(b) to 15.6	0.0(ab) to 11.5	12.6(b) to 27.1	Lauer et al. (1973). a. Combined effects: mechanical and thermal. Discharge temp. averaged 31.1°C. b. Average of two replicates Significant differences in mortality found at two stations within discharge structure. 100% survival found in one sample at one station under combined effects. c. Salinity at site ranges from 0 °/∘∘ to 7 °/∘∘.

Table 1a (Continued)

SPECIES/ORGANISMS & LIFE STAGE	GEOGRAPHIC LOCATION & PLANT DESIGNATION	EFFECTS % Mortality of Entrained Organisms					REFERENCES & REMARKS
		Physical	Thermal	Chemical	Other	Total	
Acartia tonsa (Copepod) adult	Turkey Point, Biscayne Bay, FL	-	-	-	-	42.3(a) to 92.3	Prager et al. (1970) a. Passage time, intake through pumps and condenser to discharge: approx. 1-2 min.
Acartia tonsa (Copepod) adult	Morgantown, Potomac River, MD	-	-	39.8	0.0(a) to 5.1	46.11(b)	Heinle, Millsaps & Lawson (1974) a. Combined effects: mechanical & thermal. b. Data from station at upper end of discharge canal. Data from a station at end of canal indicates significantly higher survival. This appears to be a sampling artifact. Dead organism apparently settle to bottom of canal before reaching end station.
Acartia tonsa (Copepod) egg & early nauplii	Chalk Point Patuxent River, MD	0(a)	-	-(b)	-	0.0(b) to 93.7	Heinle (1969) a. Differences noted in concentrations of surviving animals apparently are within expected natural variation.

Table 1a (Continued)

							b. Mortality derived from differences in survival between intake and effluent, based on mean of 5 or 6 samples. Range from 3 sets of experiments in 1964, 1965 and 1966 - 93.7%, 72.4% and 0% mortality, respectively. Author indicates observed difference may be due to intermittent chlorination.
Acartia tonsa (Copepod) adult	Turkey Point Biscayne Bay, FL	-	-	-	-	37.0(a)	Prager et al. (1971) a. Samples taken at intake and 200 yds from end of discharge canal. Passage approximately two hours.
Balanus sp. (Barnacle) cold temperate, Atlantic spring, March-May larvae	Long Island Sound Northport, NY	-	-	-	-	0.0(ab) to 19.8	Williams et al. (1971) a. Sampled at end of discharge canal. Exposure time <10 min. b. Three observations, range: 0-19.8% mortality.

Table 1a (Continued)

SPECIES/ORGANISMS & LIFE STAGE	GEOGRAPHIC LOCATION & PLANT DESIGNATION	EFFECTS % Mortality of Entrained Organisms					REFERENCES & REMARKS
		Physical	Thermal	Chemical	Other	Total	
Canuella canadensis (Copepod) eggs & early nauplii	Chalk Point, Patuxent River, MD	0.0(a)	-	-	-	0.0(b) to 58.6	Heinle (1969) a. Differences noted in concentrations of surviving animals apparently are within expected natural variations (author). b. Mortality figures derived from differences in survival between intake & effluent based on mean of 5 or 6 samples. Range, from 3 sets of experiments in 1964, 1965, and 1966--0.0%, 58.6%, and 7.32%, respectively. Author, indicates differences may be due to intermittent chlorination.
Eurytemora affinis (Copepod) adult	Morgantown, Potomac River, Md	-	-	24.2	0(a)	24.2	Heinle et al. (1974) a. Combined effects: mechanical & thermal.
Gammarus sp. (Amphipod) adult	Indian Point (a) Hudson River, NY	0-1.0(b)	0.3(b) to 4.9	13.2(b) to 40.1	-	18.5(b) to 48.0	Ginn, Waller & Haver (1974) a. Salinity at site ranges from 0 °/₀₀ to 7.0 °/₀₀. b. Mean percentage alive at intake compared to discharge. Numbers represent an approximation.

Table 1a (Continued)

Labidocera aestiva *(Copepod)* *adult*	*Turkey Point* *Biscayne Bay, FL*	-	-	-	-	*30.0(ab)* *to* *55.6*	*Prager et al. (1970)* *a. Passage time, intake through pumps and condenser to discharge: approx. 1-2 min.* *b. Data on <100 individuals.*
Monoculodes edwardsi *(Amphipod)* *adult*	*Indian Point (d)* *Hudson River, NY*	-	-	*22.9(ab)*	*4.4(ac)*	*27.3 (ab)*	*Lauer et al. (1973)* *a. Average of mean at 2 stations.* *b. No control samples taken for chlorine effects. Assumed control mortality 5.4%.* *c. Combined effects: mechanical & thermal. Discharge temp. 16.4-21.7°C.* *d. Salinity at site ranges from 0 °/oo to 7 °/oo.*
Neomysis americana *(Opossum shrimp)* *adult*	*Indian Point (d)* *Hudson River, NY*	-	-	*0.0 (ab)* *to* *26.4*	*6.4 (ac)* *to* *33.3*	*32.7(ab)* *to* *33.3*	*Lauer et al. (1973)* *a. Average of mean at 2 stations.* *b. No control samples taken for chlorine effects. Assumed control mortality 5.4 to 14.6%. depending on date of sample.* *c. Combined effects: mechanical & thermal. Discharge temp. 19.7 to 33.3. Highest mortality at highest temperature.* *d. Salinity at site ranges from 0 °/oo to 7 °/oo.*

Table 1a (Continued)

SPECIES/ORGANISMS & LIFE STAGE	GEOGRAPHIC LOCATION & PLANT DESIGNATION	EFFECTS % Mortality of Entrained Organisms					REFERENCES & REMARKS
		Physical	Thermal	Chemical	Other	Total	
Oithona nana (Copepod) adult	Turkey Point, Biscayne Bay, FL	-	-	-	-	53.6(a) to 96.3	Prager et al. (1970) a. Passage time intake through pumps & condenser to discharge: approx. 1-2 min.
Oithona brevicornis (Copepod) eggs & early nauplii	Chalk Point, Patuxent River, MD	-	-	-	-	43.2(ab) to 72.3	Heinle (1969) a. Mortality figures derived from differences in survival between intake & effluent based on mean of 5 or 6 samples. Range is from 2 sets of experiments in 1964 and 1966: 72.3 and 43.2%, respectively. b. Data on <100 individuals.
Oithona spp. (Copepod) adult	Turkey Point, Biscayne Bay, FL	-	-	-	-	31.0(a)	Prager et al. (1971) a. Samples taken at intake and 183m from end of discharge canal. Passage time approximately 2 hrs.
Paracalanus crassirostris (Copepod) adult	Turkey Point, Biscayne Bay, FL	-	-	-	-	41.7(a) to 100.0	Prager et al. (1970) a. Passage time, intake through pumps and condenser to discharge approximately 1-2 min.

Paracalanus crassirostris *(Copepod)* *adult*	*Turkey Point, Biscayne Bay, FL*	-	-	-	-	*70.0(a)*	*Prager et al. (1971)* *a. Samples taken at intake and 183m from end of discharge canal. Passage time approximately 2 hrs.*
Sagitta *sp.* *(Arrow worm)* *adult*	*Turkey Point, Biscayne Bay, FL*	-	-	-	-	*100(ab)*	*Prager et al. (1970)* *a. Passage time, intake through pumps & condenser to discharge: approx. 1-2 min.* *b. Data based on <100 individuals.*
Scottolana canadensis *(Copepod)* *adult*	*Morgantown, Potomac River, MD*	-	-	*25.7*	*0(a)*	*25.7*	*Heinle et al. (1974)* *a. Combined effects: mechanical and thermal.*
Tortanus aetocaudatus *(Copepod)* *adult*	*Turkey Point, Biscayne Bay, FL*	-	-	-	-	*55.4(ab)*	*Prager et al. (1970)* *a. Passage time, intake through pumps & condenser to discharge: approx. 1-2 min.* *b. Average mortality-9 samples.*
Tortanus aetocaudatus *adult*	*Turkey Point, Biscayne Bay, FL*	-	-	-	-	*7.0(a)*	*Prager et al. (1971)* *a. Samples taken at intake & 183m from end of discharge canal. Passage time approximately 2 hrs.*

Table 1a (Continued)

SPECIES/ORGANISMS & LIFE STAGE	GEOGRAPHIC LOCATION & PLANT DESIGNATION	EFFECTS % Mortality of Entrained Organisms					REFERENCES & REMARKS
		Physical	Thermal	Chemical	Other	Total	
"Barnacle" nauplii	Millstone Point Long Island Sound, Waterford, CT	-	-	-	48.0(a) to 60.0	95.0	U.S. Atomic Energy Commission Directorate of Licensing, Docket No. 50-423, Feb. 1974 a. Combined effects: mechanical & thermal.
"Bivalve" larvae	Turkey Point, Biscayne Bay, FL	-	-	-	-	42.0(a) to 100	Prager et al. (1970) a. Passage time, intake through pumps and condenser to discharge: approx. 1-2 min.
"Copepods" (cold temperate, Atlantic coast, spring species, March-May)	Long Island Sound, Northport, NY	-	-	-	-	29(ab)	Suchanek & Grossman (1971) a. Sampled at end of discharge canal. Exposure time <10 min. b. Average of 3 observations, range: 0 to 66.7% mortality.
"Copepods" (cold temperate, Atlantic coast, summer species, July-August)	Long Island Sound Northport, NY	-	-	-	-	52.5(ab)	Suchanek & Grossman (1971) a. Sampled at discharge to seal pit after condenser passage. b. Average of 9 observations. Range: 0 to 100% mortality. Highest mortalities at temp. >34°C.

Table 1a (Continued)

"Copepods" nauplii	Turkey Point, Biscayne Bay, FL	-	-	-	-	20.0(a) to 77.4	Prager et al. (1970) a. Passage time, intake through pumps and condenser to discharge: approx. 1-2 min.
"Copepods" "Larger" adults >0.333 mm	Millstone Point Long Island Sound, Waterford, CT	48-71(a)	-	-	-	70.0(b)	Carpenter, Peck & Anderson (1974) a. Combined effects: thermal & mechanical at 15.8°C. ΔT b. Average percentage of copepod "Lost". Samples at plant intake compared to samples at discharge from effluent pond. Passage through plant-22 min. Retention in pond 6-9 hrs.
"Total Copepods" (>11 species) adult	Millstone Point Long Island Sound Waterford, CT	-	-	-	12.0(a) to 75.0	65.0 to 99.0	U.S. Atomic Energy Commission Directorate of Licensing, Docket No. 50-423, Feb. 1974 a. Combined effects: mechanical and thermal.
"Crab" zoea	Turkey Point, Biscayne Bay, FL	-	-	-	-	0.0(ab) to 42.9	Prager al al. (1970) a. Passage time, intake through pumps & condenser to discharge: approx. 1-2 min. b. Data on <100 individuals.
"Crab" larvae	Long Island Sound, Northport, NY	-	-	-	-	83.7(ab)	Williams et al. (1971) a. Sampled at discharge to seal pit after condenser passage. b. Average of 7 observations, Range: 0 to 100% mortality Highest mortality at temp. >34°C. Data on <100 individuals.

Table 1a (Continued)

SPECIES/ORGANISMS & LIFE STAGE	GEOGRAPHIC LOCATION & PLANT DESIGNATION	EFFECTS % Mortality of Entrained Organisms					REFERENCES & REMARKS
		Physical	Thermal	Chemical	Other	Total	
"Gastropod" larvae	Turkey Point, Biscayne Bay, FL	-	-	-	-	0.0(a) to 85.0	Prager et al. (1970 a. Passage time, intake through pumps & condenser to discharge: approx. 1-2 min.
"Harpacticoids" adult	Turkey Point, Biscayne Bay, FL	-	-	-	-	47.9(ab)	Prager et al. (1970) a. Passage time intake through pumps and condenser to discharge: approx. 1-2 min. b. Average of 9 samples.
"Harpacticoids" adult	Turkey Point, Buscayne Bay, FL	-	-	-	-	0(ab)	Prager et al. (1971) a. Sample taken at intake and 183m from end of discharge canal. Passage time approximately 2 hrs. b. 168 individuals per m^3 at intake. 16 individuals per m^3 near end of discharge canal. Apparent sinking of killed copepods caused discrepancies in concentrations.
"Polychaete" larvae	Millstone Point, Long Island Sound, Waterford, CT	-	-	-	90.0(a)	92.0	U.S. Atomic Energy Commission Directorate of Licensing, Docket No. 50-423, Feb. 1974. a. Combined effects: mechanical & thermal.

Table 1a (Continued)

"Polychaete" larvae	Turkey Point, Biscayne Bay, FL	-	-	-	-	66.7(ab) to 100	Prager et al. (1970) a. Passage time, intake through pumps and condenser to discharge: approx. 1-2 min. b. Data on <100 individuals.
"Polychaetes"	Long Island Sound Northport, NY	-	-	-	-	11.9(ab)	Williams et al. (1971) a. Sampled at discharge to seal pit after condenser passage. b. Average of seven observations. Range was 0 to 50% mortality. Highest mortality at temperature >34°C. Data on <100 individuals.
"Shrimp" larvae	Turkey Point, Biscayne Bay, FL	-	-	-	-	50.0(ab) to 62.5	Prager et al. (1970) a. Passage time, intake through pumps and condenser to discharge: approx. 1-2 min. b. Data on <100 individuals.
"Zooplankton" (mixed) (including Calanoid copepods, Harpacticoid copepods, invertebrate larvae, and others) eggs, larvae & adults	Three Sites--Baytown, Texas; Morgantown, MD, and Brayton Point, MA	-	-	-	-	20(a) to 100	Lackie (1972-73) a. Mortality varies with size and time after exposure. Generally larger organisms had higher mortality and increased mortality occurred up to 7 days after exposure.

Table 1a (Continued)

SPECIES/ORGANISMS & LIFE STAGE	GEOGRAPHIC LOCATION & PLANT DESIGNATION	EFFECTS % Mortality of Entrained Organisms					REFERENCES & REMARKS
		Physical	Thermal	Chemical	Other	Total	
Zooplankton (mixed, including barnacles nauplii, Acartia sp. nauplii and copepodites, harparticoides, calanus nauplii)	Humbolt Bay, CA	-	-	-	0(a) to 13.0	-	Icanberry (1973) a. Range of monthly mortalities. August 1971 to July 1972. Chlorination not indicated during sampling period.
	Potrero, CA	<1.5	9.7	65.4(b)	0(a) to 5.8	72.3(b)	b. Chlorine residual 0.4 to 0.5 pp.
	Moss Landing, CA	-	-	-	0.8(a) to 38.3	-	Consult reference for tabulation of individual species mortalities at each site. Relative mortalities vary from site to site.
	Morro Bay, CA	-	-	-	0.9(a) to 29.4	-	
Phytoplankton (diatoms, five species)	Haynes-Alamitos Generating Complex Long Beach, CA	-	-	-	-	29.2(a) to 89.6	Briand (1975) a. Mortality principally due to thermal stress (author)
Phytoplankton (dinoflagellates six species)	Haynes-Alimitos Generating Complex Long Beach, CA	-	-	-	-	12.4(a) to 49.4	Briand (1975) a. Mortality principally due to thermal stress (author)

REFERENCES

Austin, H. M., J. Dickinson, and C. Hickey. 1973. An ecological study of ichthyoplankton at the Northport power station, Long Island, New York. New York Ocean Sci. Lab., Montauk, N.Y. Rept. to Long Island Lighting Co.

Briand, F. J-P. 1975. Effects of power plant cooling systems on marine phytoplankton. Mar. Biol. 33:135-146.

Carpenter, E. J., B. B. Peck, and S. J. Anderson. 1974. Survival of copepods passing through a nuclear power station on northeastern Long Island Sound, USA. Mar. Biol. 24:49-55.

Ginn, T. C., W. T. Waller, G. J. Lauer. 1974. The effects of power plant condenser cooling water entrainment on the amphipod, *Gammarus* sp. Water Res. 8:1-9.

Heinle, D. R. 1969. Temperature and zooplankton. Chesapeake Sci. 10:186-209.

Heinle, D. R., H. S. Millsaps, Jr., and J. K. Lawson. 1974. Zooplankton investigations at the Morgantown power plant. University of Maryland, Natural Resources Institute, Ref. No. 74-10.

Icanberry, J. W. 1973. Zooplankton survival in cooling water systems of four California coastal power plants, August 1971-July 1972. Presented at 36th Annual Meeting ASLO, Univ. of Utah, Salt Lake City, June 10-14, 1973.

Lackie, N. F. 1973. U.S. EPA National Marine Water Quality Laboratory, Narragansett, R.I. (unpublished data).

Lauer, G. J., W. T. Waller, D. W. Bath, W. Meeks, R. Heffner, T. Ginn, L. Zubarik, P. Bibko, and P. C. Storm. 1973. Entrainment studies on Hudson River organisms. New York University Medical Center, Laboratory for Environmental Studies Report, Nov. 1973.

Marcy, B. C., Jr. 1973. Vulnerability and survival of young Connecticut River fish entrained at a nuclear power plant. J. Fish. Res. Board Can. 30:1195-1203.

Nawrocki, S. S. 1977. A study of fish abundance in Niantic Bay with particular reference to the Millstone Point nuclear power plant. M.S. Thesis, Univ. of Connecticut, Storrs (unpublished).

Prager, J. C., E. W. Davey, J. H. Gentile, W. Gonzlalez, R. L. Steele, M. D. Lair, C. I. Weber, and S. L. Bargo. 1970. A study of Biscayne Bay plankton affected by the Turkey Point thermoelectric generating plant during July and August 1970. U.S. EPA, National Marine Water Quality Laboratory Spec. Publ., Nov. 1970.

Prager, J. C., R. L. Steele, W. Gonzlalez, R. Johnson, and R. L. Highland. 1971. Survey of benthic microbiota and zooplankton conditions near Florida Power and Light Company's Turkey Point plant, August 23-27, 1971. U.S. EPA National Marine Water Quality Laboratory, Spec. Publ., Sept. 1971.

Suchanek, T. H., Jr. and C. Grossman. 1971. Viability of zooplankton. Pages 61-83 *in* Williams, G. C., J. B. Mitton, T. H. Suchanek, Jr., N. Gebelein, C. Grossman, J. Pearce, J. Young, C. E. Taylor, R. Mulstay, and C. D. Hardy. Studies on the effects of a steam electric generating plant on the marine environment at Northport, New York. Technical Report 9 of the Marine Sciences Research Center, State University of New York.

U.S. Atomic Energy Commission. 1974. Final environmental statement. Millstone Point nuclear power station, Unit 3. Licensing Docket No. 50-423.

Williams, G. C., J. B. Mitton, T. H. Suchanek, Jr., N. Gebelein, C. Grossman, J. Pearce, J. Young, C. E. Taylor, R. Mulstay, and C. D. Hardy. 1971. Studies on the effects of a steam electric generating plant on the marine environment at Northport, New York. Technical Report 9 of the Marine Sciences Research Center, State University of New York.

CHAPTER 6. ON SELECTING THE EXCESS TEMPERATURE TO MINIMIZE THE ENTRAINMENT MORTALITY RATE

THE COMMITTEE ON ENTRAINMENT

TABLE OF CONTENTS

I. INTRODUCTION

In selecting an excess temperature at which to operate a power plant cooling system it has been customary to consider only *thermal stresses* and to use the ratio of the number of organisms killed to the number of organisms entrained. This frequently leads to the selection of a low excess temperature, ΔT, which, in turn, requires a large volume flow of cooling water. When mortalities due to *physical* and *chemical stresses* are included and the total number of entrained organisms killed is taken as the measure of the environmental damage, it becomes evident that the choice of a low excess temperature is seldom, if ever, best.

The fundamental concept is that the lower the excess temperature to be maintained, the greater must be the volume rate at which water is taken into the system. More organisms are entrained and exposed, not only to the thermal stress, but to physical and chemical stresses as well.

Each power plant draws its cooling water from a different environment which changes with space and time. Each plant operates according to its own doctrine. In spite of these individual differences there are some general characteristics which hold for all.

(1) The smaller the intake volume rate, the smaller is the number of organisms entrained.

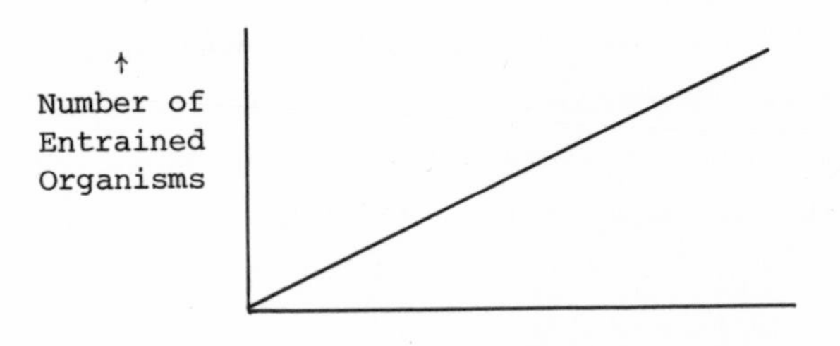

Fig. 1. Variation of number of entrained organisms with intake volume flow rate.

(2) The smaller the intake rate, the larger is the excess temperature ΔT.

(3) The lower the excess temperatures, the lower is the thermal mortality. For any initial intake water

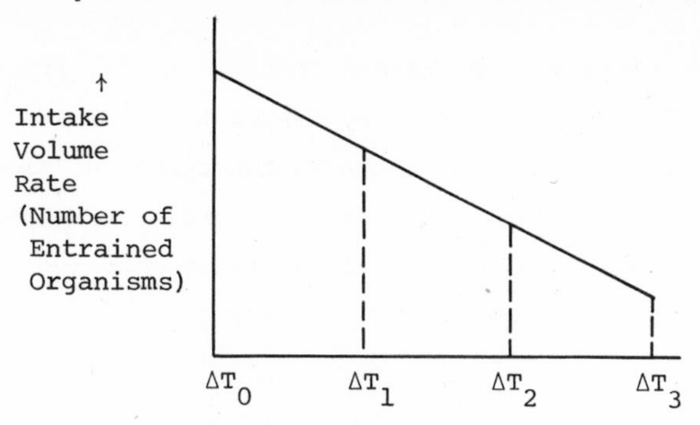

Fig. 2. Variation of intake volume flow rate (and number of entrained organisms) with excess temperature (ΔT).

temperature and any operating range (ΔT_0, ΔT_3) there is a subrange of excess temperature (ΔT_0, ΔT_1) which the entrained organisms can tolerate with little or no damage. There is also a subrange of high excess temperatures (ΔT_2, ΔT_3) in which few, if any, organisms survive. We have:

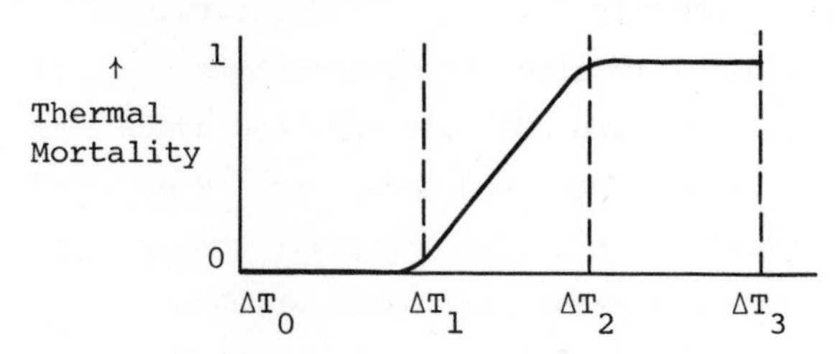

Fig. 3. Variation of number of entrained organisms killed by thermal stress as a function of excess temperature (ΔT).

(4) Even if there were no thermal changes in the system organisms would be killed by impact in pumps and ducts and by the physical stresses in the turbulent flows. The slower the flow, the less the damage from these physical stresses. Since slower flows correspond to higher excess temperatures we have:

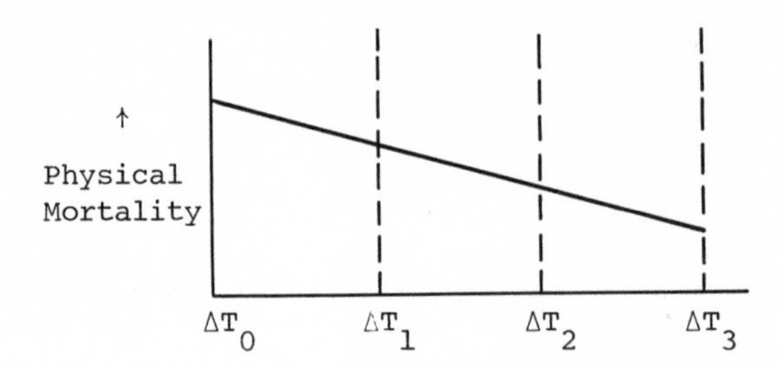

Fig. 4. Variation of number of entrained organisms killed by physical stress as a function of excess temperature (ΔT). With increasing excess temperature the volume flow rate decreases and with it the number of entrained organisms.

(5) Chemical mortality is the result of the injection of biocides, usually chlorine, to discourage the attachment of fouling organisms within the system. These biocides are not selective and are a chemical stress for all entrained organisms. Many plants apply biocides in pulses which are completely lethal during the injection period but which are diluted to less toxic levels between pulses. At higher temperatures biocides are more efficient so that less frequent and less intense chlorination are needed. Further, the residual lethal effect between pulses dies away more rapidly. Experience indicates that mortality from chemical stress decreases with

increasing excess temperature. The real reason should be obvious. At higher excess temperatures the reduced flow entrains fewer organisms that can be attacked chemically.

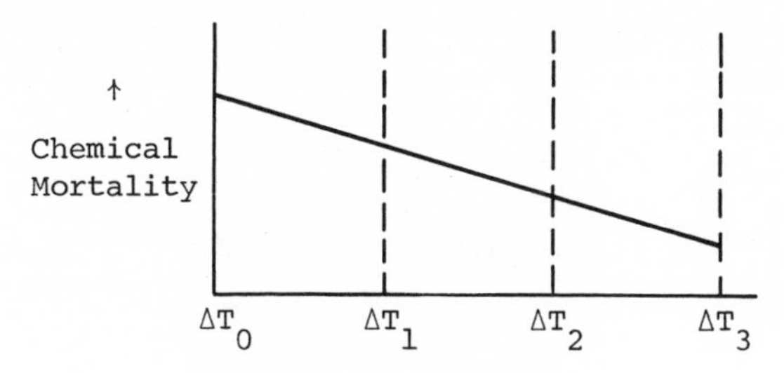

Fig. 5. Variation of number of entrained organisms killed by chemical stress with excess temperature (ΔT).

In order to find a criterion for the selection of an operating ΔT least damaging to the biota it is necessary to assemble Figs. 2 through 5 in a single model.

II. THE ENTRAINMENT MORTALITY RATE AS A FUNCTION OF EXCESS TEMPERATURE

The measure of the damage done to the biota by a cooling system will be the *entrainment mortality rate R* defined as the number of organisms killed per unit time. Contrast this to the often used *entrainment mortality fraction f* which is defined as the ratio of the number of organisms killed to the number entrained in the system. Note particularly that, if one wants to reduce the damage to the biota, it is R, not f, which must be minimized.

Let E be the rate at which organisms are entrained by the system. Then

$$R = fE \quad . \tag{1}$$

Let η be the density of organisms in the water exposed to the intake and Q be the volume rate of intake. Then, in terms of Q, the entrainment rate E is

$$E = \eta Q \quad . \tag{2}$$

The relation between the rate at which heat is required to be expelled H and the volume flow rate of coolant Q required for any excess temperature ΔT, is given by the familiar formula:

$$Q\{\Delta T\} = (\Delta T)^{-1}(H/\rho c_p) \tag{3}$$

where $\rho \equiv$ coolant density

and $c_p \equiv$ the specific heat of the coolant.

Let

$$(\eta H/\rho c_p) \equiv K \quad . \tag{4}$$

K characterizes the plant through the necessary heat disposal H, the coolant through its density ρ and heat capacity c_p, and the biological state through the organism density η. It will vary from plant to plant and from season to season. K is necessarily positive.

Substituting Q from (2) and K from (4) in (3) we have the relation between the entrainment rate E and the excess temperature ΔT.

$$E = K(\Delta T)^{-1} \quad . \tag{5}$$

The mortality fraction has three causes: thermal, physical, and chemical. We will identify them by subscripts.

$f_t \equiv$ the part of f due solely to thermal stress,

$f_p \equiv$ the part of f due solely to physical causes,

and $f_c \equiv$ the part of f due solely to chemical stress.

If one is an optimist, the mortality fractions have corresponding survival fractions, s:

$$s \equiv 1 - f$$

$$s_t \equiv 1 - f_t$$

$$s_p \equiv 1 - f_p$$

$$s_c \equiv 1 - f_c$$

The s's and f's may, if you choose, be thought of as probabilities of life or death.

For an organism to survive passage through the system it must survive all three kinds of stresses:

$$s = s_t s_p s_c$$

or in terms of mortality fractions

$$f = 1 - (1 - f_t)(1 - f_p)(1 - f_c) \quad .$$

Expanding,

$$f = f_t + f_p + f_c - f_t f_p - f_{tc} - f_p f_c + f_t f_p f_c \ .$$

All the f's lie between 0 and 1 and not more than one of them can be as large as 1. Therefore if we may neglect the second- and third-order products in comparison with the first-order terms we may write

$$f \approx f_t + f_p + f_c \quad ,$$

without modifying the qualitative relationships of interest for this discussion. The fractional mortalities depend on the excess temperature--directly and obviously for f_t, indirectly for f_p and f_c. It will be well to indicate this functional dependence on ΔT in the usual way by writing

$$f\{\Delta T\} \approx f_t\{\Delta T\} + f_p\{\Delta T\} + f_c\{\Delta T\} \quad . \tag{6}$$

Obviously, our analysis neglects higher order interactions, e.g., death due to thermal effects which would not have been fatal had the organism not already suffered non-lethal physical damage. Such higher order considerations will be minor corrections on the first order analysis and, considering the state of our knowledge and the prospects for its improvement, unlikely to become available soon.

Substituting from (5) and (6) in (1) we have

$$R\{\Delta T\} = K(\Delta T)^{-1}(f_t\{\Delta T\} + f_p\{\Delta T\} + f_c\{\Delta T\}) \quad , \tag{7}$$

the entrainment mortality rate as a function of excess temperatures. It is the minimum value of R which determines the best operating excess temperature--the one least damaging to the biota.

III. SUBRANGES OF THE OPERATING RANGE AND THE BEHAVIOR OF THE MORTALITY FRACTIONS

We have already suggested from an inspection of the thermal mortality curve, Fig. 3, that the range of ΔT has three natural subranges. Let

$(\Delta T_0, \Delta T_3) \equiv$ the full range of excess temperatures.

$\Delta T_0 \equiv$ the smallest possible excess temperature at which it is practical to operate the system. It corresponds to the largest coolant flow and the greatest entrainment of organisms.

$\Delta T_3 \equiv$ the largest possible excess temperature at which it is practical to operate the system. It corresponds to the smallest coolant flow and the least entrainment of organisms.

$(\Delta T_0, \Delta T_1) \equiv$ the subrange of excess temperatures tolerable to entrained organisms. In this subrange $f_t \simeq 0$.

$(\Delta T_1, \Delta T_2) \equiv$ the subrange of excess temperatures in which appreciable, but not total, thermal mortality occurs.

$(\Delta T_2, \Delta T_3) \equiv$ the subrange in which thermal mortality is substantially total, $f_t \simeq 1$. In this subrange there are no survivors, f_p and f_c are irrevelant, and we may take $f_t = f = 1$.

We can consider the behavior of the mortality fractions within each subrange.

As shown in Fig. 3, the thermal mortality fraction is sigmoid.

For $\Delta T_0 \leqq \Delta T \leqq \Delta T_1$, $f_t\{\Delta T\} \simeq 0$;

for $\Delta T_1 < \Delta T < \Delta T_2$, $0 < f_t\{\Delta T\} < 1$;

and for $\Delta T_2 \leqq \Delta T \leqq \Delta T_3$, $f_t\{\Delta T\} \simeq 1$.

Over $(\Delta T_0, \Delta T_1)$ and $(\Delta T_2, \Delta T_3)$, $f_t\{\Delta T\}$ is substantially constant at either 0 or 1. Over $(\Delta T_1, \Delta T_2)$ it is monotonic increasing.

As shown in Fig. 4, the physical mortality fraction, $f_p\{\Delta T\}$, decreases over the entire range $(\Delta T_0, \Delta T_3)$. Since the higher excess temperatures correspond to lower flow rates and slower speeds, less physical damage occurs. $f_p\{\Delta T\}$ is at least monotone decreasing.

Figure 5 suggests that the chemical mortality fraction, $f_c\{\Delta T\}$, also decreases over the whole range $(\Delta T_0, \Delta T_3)$. The way biocides are frequently applied to systems must be considered. It is customary to inject biocide into a system once daily for some length of time, to be specific say for one hour. During injection the chemical is fatal to nearly all the organisms present. If it were not, it could hardly be called a "biocide." During the remaining 23 hr the chemical is strongly diluted and its toxic effects are weak. Thus, the chemical mortality considered over a day varies with the number of organisms directly exposed to the biocide pulse. At high excess temperatures volume flow rates are low and so also is the number of organisms entrained. At low excess temperatures the volume flow rates are high and so is the number of organisms entrained. However, since f_c is a fraction of the entrained organisms it can, at most, be constant on this argument. On the other hand, fewer organisms are exposed to chemical stress at high ΔT than at low ΔT which is a definite gain. What evidence we have suggest that f_c is a monotone decreasing function of ΔT.

The factor $(\Delta T)^{-1}$ is monotonic decreasing over the entire range $(\Delta T_0, \Delta T_3)$.

IV. BEHAVIOR OF THE ENTRAINMENT MORTALITY RATE ON THE SUBRANGES

Consider $R\{\Delta T\}$ on $(\Delta T_0, \Delta T_1)$. In this subrange f_t is effectively 0 so that (7) becomes

$$R\{\Delta T\} = K(\Delta T)^{-1}(f_p\{\Delta T\} + f_c\{\Delta T\}) \quad .$$

$f_p\{\Delta T\}$ is monotone decreasing.

$f_c\{\Delta T\}$ is monotone decreasing or, at worst, constant.

Therefore, their sum is monotone decreasing.

$(\Delta T)^{-1}$ is monotonic decreasing.

Therefore, $R\{\Delta T\}$, which is the product, is monotonic decreasing.

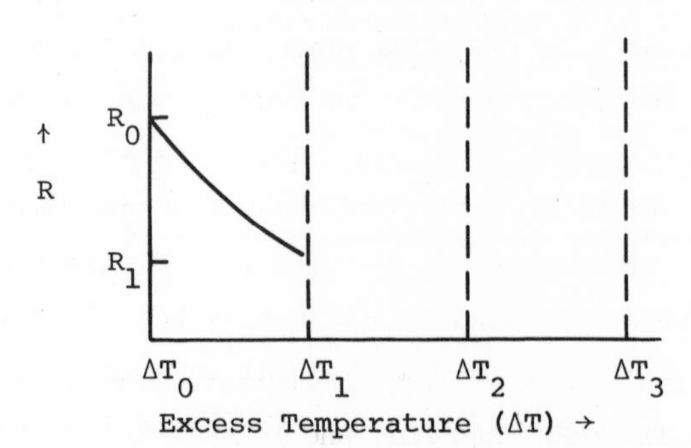

Fig. 6. Variation of the entrainment mortality rate, R, with excess temperature (ΔT) over the subrange ΔT_0 to ΔT_1. ΔT_0 is the smallest excess temperature at which it is practical to operate a plant. ΔT_1 is the maximum temperature for which mortality from thermal stress is zero.

From this simple argument it is obvious that within $(\Delta T_0, \Delta T_1)$ the most damaging excess temperature at which to run the system is the lowest possible excess temperature ΔT_0. Any other will do less harm and the highest on the subrange, ΔT_1, will do the least.

The situation on $(\Delta T_2, \Delta T_3)$ is equally clear. On this range f_t is substantially constant equal to 1 while f_p and f_c are irrelevant. Equation (7) becomes

$$R\{\Delta T\} = K(\Delta T)^{-1}$$

which is monotonic decreasing.

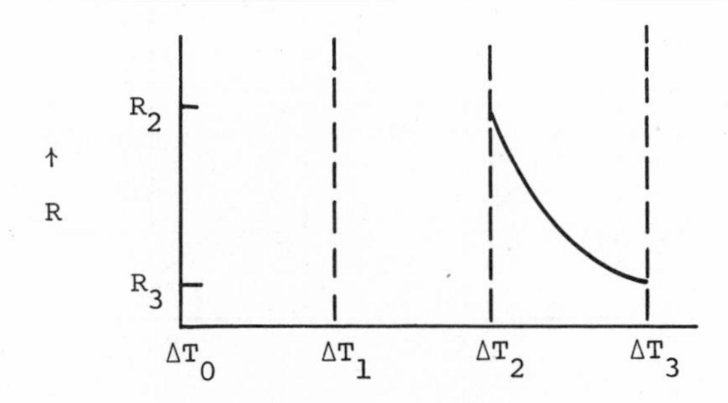

Fig. 7. Variation of the entrainment mortality rate, R, with excess temperature (ΔT) over the subrange ΔT_2 to ΔT_3. ΔT_2 to ΔT_3 is the subrange over which thermal mortality is substantially total. ΔT_3 is the largest excess temperature at which it is practical to operate a plant.

As before, on the interval $(\Delta T_2, \Delta T_3)$ the lowest excess temperature is the most damaging and the highest least.

If only $(\Delta T_0, \Delta T_1)$ is considered, the choice will be ΔT_1. If only $(\Delta T_2, \Delta T_3)$ is considered, the choice will be ΔT_3. If both are considered, further information is needed since we may have $R_1 \lesseqgtr R_3$. In any case, all other excess temperatures on these intervals are worse and, in particular, ΔT_0 and ΔT_2 are very bad. The moral is that:

> IF YOU WANT TO DO AS LITTLE HARM AS POSSIBLE, DON'T EVER OPERATE AT A SMALL ΔT.

The situation on the central subrange (ΔT_1, ΔT_2) can not be had by a simple argument. There, f_t is monotone increasing while f_p and f_c are monotone decreasing. Their sum may be neither. If the curve positions are as sketched in Figs. 6 and 7, then there must be at least one relative minimum and one relative maximum in (ΔT_1, ΔT_2). It is unlikely that there will be more.

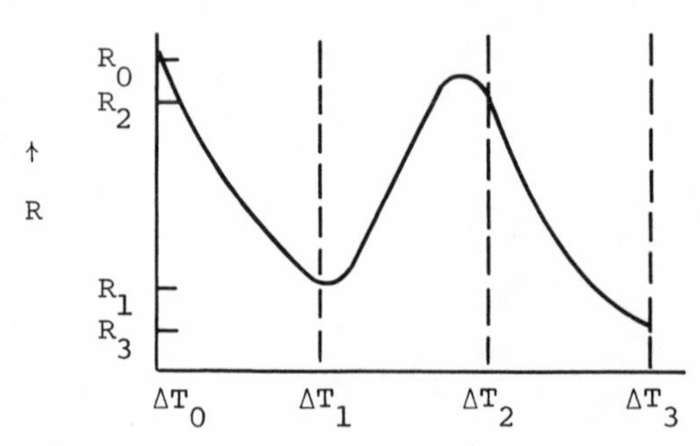

Fig. 8. Possible variation of the entrainment mortality rate, R, with excess temperature over the entire range of ΔT, ΔT_0 to ΔT_3.

On the other hand, Figs. 9 and 10 are also possible, among others. Only the case with a minimum on (ΔT_1, ΔT_2) offers any hope for an operation excess temperature less harmful than ΔT_1 or ΔT_3. If a minimum in R exists at $\Delta T_{R_{min}}$, it will certainly be an improvement over ΔT_1. It may or may not be an improvement on ΔT_3.

As our knowledge increases we can hope to improve our understanding of $R\{\Delta T\}$, particularly on (ΔT_1, ΔT_2). As an immediate practical matter we can reduce damage to the biota by concentrating on the choice between $R\{\Delta T_1\} \equiv R_1$ and $R\{\Delta T_3\} \equiv R_3$.

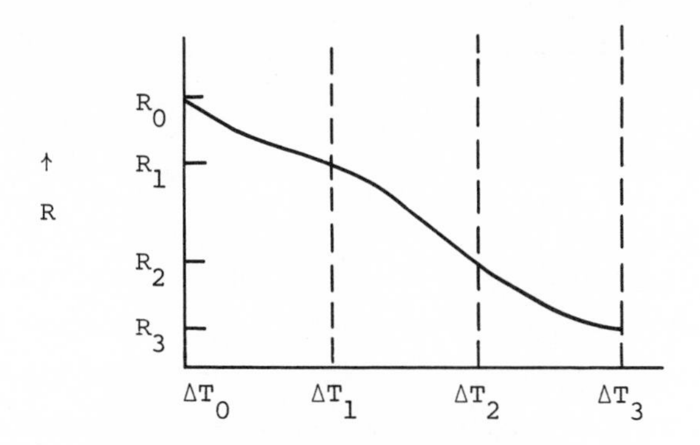

Excess Temperature (ΔT) →

Fig. 9. Possible variation of the entrainment mortality rate, R, with excess temperature over the entire range of ΔT, ΔT_0 to ΔT_3.

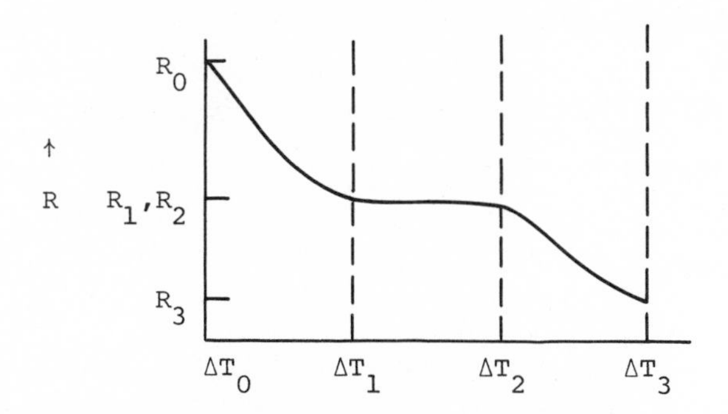

Excess Temperature (ΔT) →

Fig. 10. Possible variation of the entrainment mortality rate, R, with excess temperature over the entire range of ΔT, ΔT_0 to ΔT_3.

R_1 is certainly an improvement over $R_0 \equiv R\{\Delta T_0\}$ and will be the chosen excess temperature in the absence of other information. If $R_3 \equiv R\{\Delta T_3\}$ is also known, then the minimum of R_1, R_3 will determine the choice between ΔT_1 and ΔT_3.

In brief, only three excess temperatures need be considered: ΔT_1, $\Delta T_{R_{min}}$, and ΔT_3. Of these, use the one whose R-value is least. If $\Delta T_{R_{min}}$ is near ΔT_1 while R_{min} is near R_1, inclusion of $\Delta T_{R_{min}}$ in your considerations will not improve much on ΔT_1.

V. THE CRITERION

Application of this analysis in the absence of information calls for using the highest excess temperature at which it is possible to operate the system, ΔT_3. In such circumstances it is true that nearly all organisms passing through the system will die but the number entrained would be greatly reduced thus, one hopes, minimizing the damage to the entire population at risk.

The next higher level of application requires the determination of ΔT_1. The value of ΔT_2, the excess temperature, at which thermal mortality reaches 100% for the first time, is irrelevant unless the highest practical excess temperature, ΔT_3 is less than ΔT_2; which is unlikely. Once you begin killing all the organisms involved the only way to reduce the damage is to reduce their numbers.

To determine ΔT_1 experimental thermal resistance curves, similar to toxicity curves, are needed. They can be expected to be different for different initial (ambient) temperatures, excess temperatures, and durations of exposure before return to ambient temperature as well as for different species and stages of development.

If only organisms of a single species were entrained and that species were the same at each plant, the preparation of thermal tolerance curves would be comparatively easy. As it is, the species composition of the population at risk of entrainment differs from one plant to another and from season to season at any one plant. It is recommended that the thermal tolerance curves be determined for the *Representative Important Species* (RIS). A species may be important for one of three reasons:

(1) It may be commercially or recreationally valuable.

(2) It may be a critical link in the life-web of the region.

(3) It may be a sensitive indicator of the thermal responses of a number of other species.

It would be reasonable to choose for ΔT_1 that excess temperature at which *any* important *sensitive* species developed 10% mortality; the assumption being that few other species would suffer mortalities nearly as high and that most would experience little or no mortality. Other, equally reasonable, suggestions could be made.

Whatever method of choosing ΔT_1 is adopted, it would seem wise to concentrate first on studies of important species common to many plant sites and suspected of having low thermal tolerances.

Once ΔT_1 is known, equation (7) becomes

$$R_1 \equiv R\{\Delta T_1\} = K(\Delta T_1)^{-1}(F_{p,c}\{\Delta T_1\})$$

where

$$F_{p,c}\{\Delta T_1\} \equiv f_p\{\Delta T_1\} + f_c\{\Delta T_1\} \quad .$$

If the organism density at the intake, the heat to be disposed of, and the density and specific heat of the coolant are known, K is known so that the entrainment mortality rate R_1 is known up to a multiplicative factor, $F_{p,c}\{\Delta T_1\}$, dependent on the physical and chemical stresses.

Equation (5) gives us the entrainment rate E as a function of the *excess temperature*. At ΔT_3 all entrained organisms are killed so that E is R and

$$R_3 \equiv R\{\Delta T_3\} \equiv E\{\Delta T_3\} = K(\Delta T_3)^{-1} \quad .$$

Since we want to compare R_1 with R_3 and since they have the common factor, K, a ratio will be useful. Form

$$\begin{aligned} R_1/R_3 &= [K(\Delta T_1)^{-1}(F_{p,c}\{\Delta T_1\})]/[K(\Delta T_3)^{-1}] \\ &= (\Delta T_3/\Delta T_1)\ (F_{p,c}\{\Delta T_1\}) \quad . \end{aligned}$$

If $(\Delta T_3/\Delta T_1)^{-1}(F_{p,c}\{\Delta T_1\}) > 1$, $R_1 > R_3$ and ΔT_3 is used.

If $(\Delta T_3/\Delta T_1)^{-1}(F_{p,c}\{\Delta T_1\}) < 1$, $R_1 < R_3$ and ΔT_1 is used.

The ratio $\Delta T_3/\Delta T_1 > 1$ while $F_{p,c}\{\Delta T_1\}$ is certainly less than 1 and probably much less.

A simpler way to state the criterion is:

If $\Delta T_3 > F^{-1}_{p,c}\{\Delta T_1\}(\Delta T_1)$ use ΔT_3 .

If $\Delta T_3 < F^{-1}_{p,c}\{\Delta T_1\}(\Delta T_1)$ use ΔT_1 .

We may define an indifference factor by

$$\Delta T_3 = F^{-1}_{p,c}\{\Delta T_1\}(\Delta T_1) \quad .$$

For this value of $\alpha \equiv F^{-1}_{p,c}\{\Delta T_1\}$, $R_1 = R_3$ and it does not matter which excess temperature is used.

TABLE 1 The Indifference Factor

$F_{p,c}\{\Delta T_1\}$	α
0.2	5.00
0.3	3.33
0.4	2.50
0.5	2.00
0.6	1.67
0.7	1.43
0.8	1.25

What Table 1 says is that if you operate a plant at ΔT_1, the excess temperature at which thermal effects just begin to kill, and if at that temperature you are killing 20% of the entrained organisms by physical and chemical stresses then, if your highest possible operating excess temperature, ΔT_3, is more than 5 times as large as ΔT_1, you will kill fewer organisms by killing them all with heat. The bottom line says that if you kill 80% by physical or chemical stresses before temperature becomes an important killer, then ΔT_3 need be only 1.25 times ΔT_1 before you kill fewer by relying only on temperature.

For fixed heat loss requirement, H, coolant density and heat capacity, ρ and c_p, and density--and composition--of population at risk, η, the "constant" K is constant with respect to ΔT and cancels out of the ratio, $\Delta T_1/\Delta T_3$. However, $K = K(\vec{x},\tau)$, is a function of space and time. If two different plants are to be considered, or the same plant at two different seasons, then a ratio K_1/K_2 will remain. Also, the selection of ΔT_1 depends on the most thermally sensitive Representative Important Species at risk, which may well change seasonally and with plant location. Out of nothing, nothing. Except for the simplest comparisons we should know something about K.

It would clearly be useful to have some experimental information about the mortality fractions due to physical and chemical stresses for the Representative Important Species when ΔT is near ΔT_1. It would appear that, unless a possible $\Delta T_{R_{min}}$ is to be taken into account, this would be sufficient for a fully informed choice. In any case, if ΔT_3 is greater than 5 times ΔT_1, ΔT_1 will be a most unlikely choice.

CHAPTER 7. CONCLUSIONS AND RECOMMENDATIONS

THE COMMITTEE ON ENTRAINMENT

TABLE OF CONTENTS

I. INTRODUCTION

All small drifting organisms, and some fairly powerful swimmers, are susceptible to being drawn into power plant intakes along with cooling water and passed through the plant (pump entrainment). Entrained organisms range from microscopic bacteria and plankton to small fish. Occasionally, when intake water velocities are high ($\geq$ 0.5 m/s) even large trout and salmon--two strong swimmers--may be trapped in intake structures and killed on the protective screens. Organisms may also be drawn into the discharge plume without passing through the plant (plume entrainment). The kinds and relative abundances of entrained organisms vary greatly from location to location and depend to a large extent on the placement and design of a plant's intake and discharge structures, and on the spatial and temporal distributions of organisms in the contributing water body.

Entrained organisms are subjected to a variety of thermal, chemical, and physical stresses during passage through a plant's cooling system. Data from power plant studies throughout the United States show that many organisms do not survive pump entrainment; mortalities range from about 1% to 100%, and a median mortality of about 30% for all trophic levels may be representative (Chapter 6). Mortality depends to a large extent on the size and fragility of the organisms; larger and more fragile organisms are more vulnerable to the physical stresses of entrainment than smaller and less fragile forms. Organisms also differ substantially in their tolerance of thermal stresses; meroplankton are usually more sensitive than other organisms and life stages.

Power plants can be viewed as selective predators that may not only reduce the abundance of vulnerable organisms but which may also disrupt community structure through selective cropping and concomitant enhancement of surviving species. Cropping by power plant entrainment is a man-induced perturbation which may be of relatively large magnitude and which should be considered as additive to other mortalities. The net effects of the cropping on the population may not be additive because of compensatory responses.

The three kinds of stresses--chemical, physical, and thermal--experienced by entrained organisms during inner-plant passage act separately and sometimes in concert. The relative importance of the stresses in determining mortality varies from plant to plant, and with time at a given plant. Thermal stresses could be the most important mortality factor during periods of high ambient temperature; chemical stresses may dominate during periods of heavy chlorination. Physical stresses which include rapid pressure changes, shear forces, turbulence, impact, and abrasion are less variable seasonally than thermal and chemical stresses and can cause high mortalities of many species throughout the year. The three classes of stresses frequently act in combination to produce the highest mortalities during the warmer months when rapid growth of fouling organisms requires relatively heavy chlorination. A critical examination of data from entrainment studies indicates that in those relatively few cases where it has been possible to apportion the probable causes of mortality among the chemical, thermal, and physical stresses, the physical stresses dominate most frequently (Chapter 5).

Entrainment losses associated with once-through (open-cycle) cooling have created considerable concern during the past decade. The concern has been directed primarily to the thermal stresses which have been perceived by many as the most important factor in causing entrainment mortality. In an attempt to mitigate any deleterious thermal effects, relatively stringent thermal regulations have sometimes been adopted. Enforcement of these regulations has not produced the desired results however, and concern over entrainment losses has intensified. Recently, closed-cycle cooling (cooling towers) has been specified as the best technology for cooling steam electric generating plants. Stringent thermal standards and construction of cooling towers have a common objective--to minimize entrainment losses--but the strategies are in a sense contradictory. Closed-cycle cooling minimizes entrainment losses by drastically reducing the volume of cooling water required and with it the number of entrained organisms. Strategies for reducing entrainment losses with once-through cooling have often been directed *not* at reducing the flow of cooling water required, but rather at decreasing the excess temperatures--a practice that increases the flow of cooling water and with it the number of entrained organisms.

The scientific and engineering expertise exist to design power plants with once-through cooling systems to minimize the total number of organisms killed by entrainment, but this is all too infrequently done. The reasons are varied but relate frequently to inadequate and inappropriate thermal standards and criteria. Once-through cooling is an acceptable mode of cooling in some environments including: open coastal areas, large lakes, large rivers, and estuaries, away from important spawning and nursery areas. In some cases, once-through cooling systems--if properly designed and operated--may be the "best" (most acceptable) method when both environmental and economic factors are weighed carefully.

II. MINIMIZING ENTRAINMENT MORTALITY

A. Site Selection

Much more attention should be directed to proper site selection than is usually done to ensure acceptable combinations of power plant design and biological value of the local environment. The effects of entrainment losses will be less if plants with once-through cooling systems are located in areas where plankton, particularly meroplankton, is relatively scarce or more tolerant to entrainment stresses than if they are sited in areas with greater biological value or where organisms are less tolerant to stress. There is a range of plant-environment combinations. At one extreme is a plant with a relatively low intake flow--low relative to the available cooling water--sited in an area of low biological value. At the other is a plant with a relatively high cooling water flow sited in an area of high biological value. Probably the first is always biologically acceptable; the second rarely, if ever. A plant with a high cooling water demand may be acceptable in an area of high biological value if entrainment losses are small either because the number of entrained organisms is small or because the survival of entrained organisms is high. Most plants operate between the two extremes and detailed evaluations are required to assess their acceptability. These studies should include, for proposed and operating plants, not only detailed site studies but also extensive--in both time and space--water body studies to assess the relative biological importance of the site to the remainder of the water body. Entrainment losses should also be documented at existing plants.

The Federal Water Pollution Control Act Amendments of 1972 [Public Law 92-500, Section 316(b)] require cooling water intake structures to reflect the best technology available for minimizing adverse environmental impact[1]. The U.S. Environmental Protection Agency (1977) provides guidance for evaluating adverse impact of cooling water systems, including organism entrainment, on the aquatic environment. A biological value-potential impact decision criteria matrix is suggested in Fig. 1 of the guidance manual, and is shown below.

	COOLING WATER FLOW (Relative to Source Water Body Segment)	
BIOLOGICAL VALUE	High	Low
High	No	Questionable
Low	Questionable	Yes

[1]See Chapter 1 for discussion of laws pertaining to environmental impact of entrainment.

Coutant (1974) suggested a six-step process that may be used to assess the impacts of entrainment:

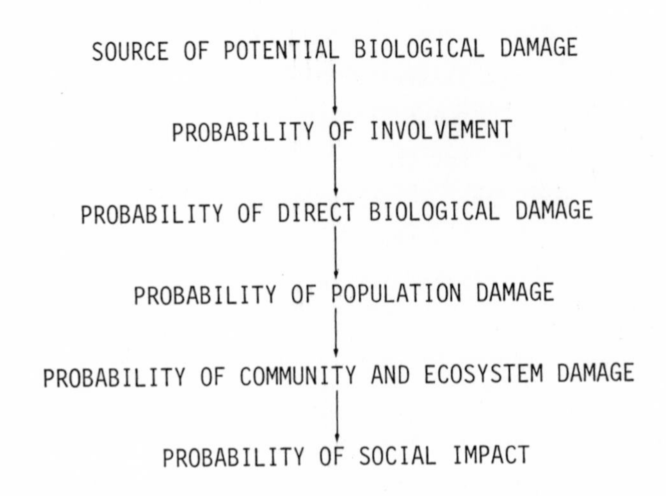

A power plant is clearly a source of potential biological damage; the extent and biological significance of that damage are determined by a set of stochastic processes which can be described only statistically. It is only through a probability distribution function that we can describe how many organisms might be "involved" with a power plant--entrained, entrapped, impinged, or otherwise affected. Once "involved," the fate of those organisms can, and indeed must, be described statistically in terms of the probability of mortality and sublethal effects. There is also some probability distribution that relates damages from entrainment to the populations of those organisms, to the community, and to the ecosystem. To properly evaluate the impact at any step in this process we must have the multivariate joint probability density function that describes that and all previous steps. It is very unlikely that any model that assumes independence will be satisfactory. There may also be social impacts resulting from biological effects on the aquatic environment.

At present, we can make adequate assessments of the probability of entrainment and of acute biological effects. Assessments of sublethal effects are less satisfactory, and diagnostic predictions of the effects of power plant cropping on populations and higher organizational levels are rarely possible. The assessment of social impacts depends not only upon biological considerations but upon a variety of economic and aesthetic factors as well. To improve our capability to assess and predict population and community effects of entrainment, we must direct more attention to water body-wide studies. Detailed site studies can not provide the basis for assessing the ecological and social significance of power plant cropping unless entrainment mortalities can be reduced to near zero levels. With respect to Coutant's six step process then, we can handle the first three steps to some degree, perhaps even adequately, but the remaining three steps are beyond present capabilities, except in rare instances. It is distressing that we continue to focus our attention on items 1-3; perhaps it is because we know how to attack them. Until population, community, and ecosystem effects can be assessed with acceptable accuracy, power plants with once-through cooling systems should be designed and operated to minimize entrainment losses. This reduces the probability that undesirable impacts of entrainment will be manifested at higher organizational levels.

B. Plant Design and Operating Criteria

After an appropriate site has been selected for a power plant, its cooling system should be designed and operated to minimize the entrainment mortality rate *R* which is defined as the number of organisms killed per unit time (Chapter 6). *R* results from three stresses: thermal, physical, and chemical. These can be identified by subscripts:

$R_t \equiv$ the number of organisms killed per unit time solely by thermal stress.

$R_p \equiv$ the number of organisms killed per unit time solely by physical stress.

$R_c \equiv$ the number of organisms killed per unit time solely by chemical stress.

If one neglects second and higher order interactions, $R = R_t + R_p + R_c$. To minimize R may, and probably would, require operating conditions different from those to minimize the entrainment mortality fraction *f* which is defined as the ratio of the number of organisms killed to the number entrained in the system (Chapter 6). The entrainment mortality fraction, f, results from three kinds of stresses: thermal, physical, and chemical. These can be identified by subscripts: f_t, f_p, and f_c. Unfortunately, present regulations focus on thermal stresses and are directed primarily at reducing f by reducing the number of organisms killed by thermal stresses alone, f_t, with little consideration for the relative importance of the mortalities from the other stresses. Reducing f may not be the best way to minimize R unless f can be reduced to perhaps a few percent. If the strategy is to reduce f by decreasing ΔT to reduce thermal stress, this can be accomplished only by increasing the flow of cooling water and therefore the total number of organisms exposed to the other stresses of entrainment. The appropriate question to ask is: x% mortality of what? Is 100% mortality of 5×10^6 entrained fish eggs per day "better" than 51% mortality of 10×10^6 fish eggs per day?

To minimize mortality--both f_t and R_t--associated with thermal stress *alone*, the appropriate procedures to follow are to: operate a plant at the lowest ΔT that is technologically possible and economically acceptable, minimize the transit time through the plant, and maximize the rate of dilution of the discharge water. This is not to imply that the suggested mode of operation would minimize the total mortality of organisms killed by all stresses. IT WOULD NOT.

To minimize mortality--both f_c and R_c--associated with chemical stresses of biocides (most commonly chlorine) *alone*, the appropriate procedures to follow are to: chlorinate intermittently, keep the level of free chlorine as low as possible, minimize the transit time through the plant, and maximize the rate of dilution of the discharge water. During periods of limited growth of fouling organisms reduction in the frequency and intensity of chlorination will also decrease unnecessary mortality from chemical stress.

To minimize mortality of entrained organisms associated with physical stresses the primary control--at the present time--must be through reducing R_p by decreasing the number of entrained organisms. The most effective way of achieving this is by decreasing the flow of cooling water. Proper design and placement of the intake can also reduce the number of pump entrained organisms. The part of the entrainment mortality fraction due to physical stress, f_p, appears to be controlled largely by the design of the circulating water pumps. The total mortality of organisms from physical stresses is directly proportional, approximately, to the flow of cooling water; doubling the flow exposes twice as many organisms to the same set of physical stresses and increases the intensity of the physical stresses. Decreasing the flow decreases the number of entrained organisms, but a higher ΔT must be used since a plant must reject heat at a fixed rate to operate efficiently.

Since up to some threshold of cooling water flow physical stresses usually cause greater mortality than thermal stresses, the most effective way of minimizing R is to decrease flow to the lowest acceptable level. This minimum flow is determined by the relative magnitudes of f_p and f_c. Setting the cooling water flow fixes the ΔT. Selection of the appropriate combination of flow and ΔT--the combination that minimizes R--must be based on the thermal resistances of Representative Important Species (RIS), and on f_p. In every case, a plant should operate either at the highest ΔT that is biologically acceptable, or at the highest ΔT that is technologically and economically feasible. A plant should never be operated at a ΔT lower than the maximum biologically permissible level; operating below this ΔT unnecessarily exposes more organisms to the other stresses of entrainment.

At any given plant the most desirable combination of ΔT and flow rate may, and probably will, change from season to season. New plants should be designed for flexible operation. The ΔT at many existing plants can be manipulated to some extent by varying the number of circulating water pumps. At many plants, the ΔT is well below the minimum desirable temperature to minimize R and should be raised. Raising the ΔT not only decreases the flow and therefore the number of entrained organisms, it has other benefits. It increases the effectiveness of biocides so that smaller amounts can be used; and it increases their rates of dissociation.

Coordinating plant activities with organism density is another approach that can help reduce entrainment losses. For example, plant shutdowns for refueling and maintenance can, at least to some extent, be scheduled to coincide with periods when the more important vulnerable organisms are most abundant. Intakes can be designed to pump from various depths and therefore draw water from zones of relatively low organism density. Many organisms migrate vertically on a day-night cycle; others, for example fish eggs, may be concentrated by physical processes. Pumping rates can also be adjusted to natural variability in the density of organisms; some variability for example diurnal variations and tidal variations are predictable. Auxiliary pumps to increase the dilution and therefore the cooling of the discharge water should rarely, if ever, be used because of the associated physical stresses.

Since ichthyoplankters are usually more sensitive than other organisms to entrainment stresses, particularly thermal and physical stresses, and since their longer regeneration times make them more vulnerable to persistent effects of a given cropping rate, protection of these organisms should ensure adequate protection of other organisms.

III. RESEARCH PRIORITIES

We have identified a number of research objectives that we believe must be attained for a significant improvement in our ability to design once-through cooling systems for minimal and predictable entrainment losses.

A. Thermal

1. Thermal Resistance Curves

Thermal resistance curves such as those shown in Chapter 2 should be developed for a variety of important organisms, particularly meroplankton and juvenile fishes. These curves should be determined for mortalities ranging from 10% to 90%.

2. Time-Excess Temperature Histories

Thermal resistance curves and engineering design criteria should be used to develop a variety of time-excess temperature exposure histories that could result in predictable and acceptable mortalities from thermal stress.

3. Verification of Predicted Mortalities

Mortalities predicted for these time-excess temperature histories should be verified in the laboratory.

4. Sublethal Effects

Sublethal effects of acute thermal shock, particularly on egg and larval stages, should be investigated. Such effects may disturb development of sensory-nervous systems, or affect morphogenesis and lead to behavioral or morphological changes that would affect survival in subsequent life stages.

B. Biocides

1. Chemistry of Chlorine

Research is needed on the chemistry of chlorine and ozone in natural waters; of particular importance are the formation of organic halogenated compounds and assessment of their carcinogenicity.

2. Tolerance Studies

Carefully controlled laboratory studies are required to determine the tolerances of a variety of organisms in different life stages to chlorine and its derivatives.

3. Effects on Behavior

Studies are required to determine the behavioral responses of a variety of organisms, particularly invertebrates, to halogenated effluents.

4. Biomagnification

Studies are needed to document the accumulation and biomagnification of halogenated organics produced by power plant chlorination.

C. Physical

1. Effects on Meroplankton

The most important problem is to determine the effects of physical stresses of entrainment on a variety of organisms, particularly meroplankton. Physical Resistance Curves should be developed for a variety of organisms in different life stages.

2. Isolating the Sites of Physical Damage

Pumps appear to cause the greatest physical damage to entrained organisms; these effects need to be verified. Alternate pump designs should be investigated and their physical effects assessed. Other sites of significant physical damage need to be identified.

3. Apportioning Physical Effects Among the Several Stresses

Studies should be done to apportion the physical effects of entrainment among the several stresses--pressure, shear, acceleration, impact, and abrasion. Recently, laboratory studies have been designed to simulate the physical and thermal stresses associated with entrainment, to separate their effects, and to identify the specific stresses (G. V. Poge et al., in preparation).

4. Size, Stage of Development and Physical Effects

More data are needed to establish the relationships between size, stage of development, and vulnerability to damage from physical stresses of entrainment.

5. Plant to Plant Variability

Assessments should be made of plant-to-plant variability of the mortalities of entrained organisms from physical stresses, and the reasons for these differences should be established.

REFERENCES

Coutant, C. C. 1974. Effect of entrainment effects. *In* L. D. Jensen, ed. Proceedings, Second Entrainment and Impingement Workshop, Johns Hopkins University, Cooling Water Research Project, Rept. No. 15.

Poje, G. V., T. C. Ginn, J. M. O'Connor. In prep. Responses of ichthyoplankton to stress simulating passage through a power plant condenser tube. *In* Proceedings, Symposium on Energy and Environmental Stresses in Aquatic Systems.

U.S. Environmental Protection Agency. 1977. Guidance for evaluating the adverse impact of cooling water intake structures on the aquatic environment. Section 316(b) P.L. 92-500, A Draft. U.S. Environmental Protection Agency, Office of Water Enforcement, Permits Division, Industrial Branch, Washington, D.C., May 1, 1977. 59 p.

APPENDIX A

This appendix consists of an engineering description of the sequence of stresses experienced by entrained organisms carried through a "typical" once-through cooling system, and a parallel account of how an old fish that survived the trip as a larva might describe the adventure to the "small fry" in the community.

PURGATORIO--TWO RATHER DIFFERENT VIEWS OF THE SAME EVENT

ROBERT E. ULANOWICZ
BLAIR KINSMAN

Descriptions of the experiences of organisms entrained in a once-through cooling system usually treat either stresses from a single cause or stresses connected with a part of the system. For example, the American Nuclear Society (1974) describes pressure changes while Coutant et al. (1975) discuss the combined thermal and physical stresses within the condenser tubes. The advantage of manageable problems encourages such partial descriptions. However, since it is our purpose to consider the total

"The year class will come to order." The portly striped bass with scarred gill plates and half his dorsal fin missing flexed his back once and snapped his jaws. The fry fluttered their tails and gave the professor the kind of nervous attention that is partly respect but principally a suspicion that they may be dinner.

"This course is Advanced Survival, S 303," the professor continued.

"Its prerequisites are Elementary Survival, S 101, and Intermediate Survival, S 202.

impact, it will be useful to follow an organism through the system.

In S 101 you have learned of the natural hazards of inner space, that bounded by the bottom and the water surface. In S 202 you have learned the more common dangers arising from the bug-eyed monsters who inhabit outer space: bloodworms which conceal barbed hooks, nets, weirs, and similar devices that descend on us through the surface. Formerly, the work on survival terminated here. However, a new and more atrocious invasion, the power plant cooling system, is multiplying and, in the opinion of the faculty, you should not be allowed to continue in your customary abysmal ignorance. Since, in my youth, I was once entrained in a cooling system--and lived--I have been appointed as your instructor.

The passage begins under ambient conditions as the organism is drawn into the water moving toward the intake. This may happen at distances as far as 100

"The most insidious thing about entrainment is that you will not know that it is happening. Everything looks and feels normal. Grazing is normal.

meters. He drifts toward the intake at a gentle 15 to 30 cm/sec and may take 5 or 10 minutes to reach the screen. Aside from distant noise he sees little out of the ordinary except for an unusual number of predators.

Temperature is normal. However, off in the distance there is a bit more noise than usual. But that's all. If at the instant you detect the noise, you swim vigorously with the noise to one side of you--not away from it--you fry just may escape. However, such evasive action in response to every slight increase in noise is impractical. For a long time there is nothing much out of the way to be seen, but you will notice that predators seem to be getting unusually plentiful. Appropriate action in the face of predators was covered in S 101. Remember the school motto: 'Edere Non Ederi.'

seconds

1 1

Conditions begin to depart from normal near the intake screens. Speeds increase slightly to 30 to 60 cm/sec. Accelerations in the turbulent flow reach approximately one-tenth of the acceleration of gravity (0.1g). The shear forces near the screen may reach 20 dynes/cm^2. Neither the

"In the last few seconds as you approach the screen the noise level goes up sharply. The meshes of the screen are large enough for you to slip through quite easily, but small enough to hang up that white perch who regards you as food. There is considerable satisfaction in seeing your pursuer

shears nor the accelerations will be damaging to most entrained organisms, although ichthyoplankton large enough to be trapped against the screen would die.

gasping his life out flattened against the screen. But you had better look sharp! You won't be able to hang around and your troubles have really begun.

20

During the 10 to 20 seconds required for the 10-meter passage from screen to pump disturbances increase abruptly. For a pump above water level the absolute pressure falls continuously from the intake. For example, an organism entrained from a depth of 10 meters experiences a drop from 2 atm to 0.3 atm. Velocities increase to 100 to 200 cm/sec. Turbulent accelerations reach 0.4 g to 1.6 g. Shears remain sublethal at 100 to 200 dynes/cm^2 near the surface of the conduit.

To prevent fouling, power plants inject pulses of biocide into the coolant. Those organisms directly exposed to an injection experience concentrations as great as 3 ppm of

21

"As you enter the conduit the most disorienting sensation will be the reports from your lateral lines. As you know, these give directional pressure signals which permit you to turn away from an attacker. In the conduit you will feel yourself completely surrounded by predator. The walls go by with a rush and you will feel yourself twisted and rolled by the current. You will have the sensation that you have surfaced from 20 meters in only a few seconds. There will be thousands and thousands of others with you and those with vacuoles will begin to explode. Then there is the burning biocide; not always but too often for comfort. If you are caught directly in one of these gas

chlorine.

attacks, you and everyone around you has bought it. In my own experience, I was fortunate enough to go through just behind a gas attack but the burns were, and still are, painful.

During the next split-second within the pump the most severe shocks occur. There is an almost instantaneous jump in pressure of about 1.5 atm. Entrainment into the boundary layer of the impeller where viscous stresses reach 10^3 to 10^4 dynes/cm^2 is a strong possibility. These stresses are an order of magnitude greater than the stresses experienced elsewhere in the system and exceed lethal levels for striped bass eggs and larvae (Morgan et al., 1976). Direct collision with the impeller will occur for some 3.5% of the entrained organisms. The impact velocity will be 1600 cm/sec, the equivalent of a fall from a height of over 15 meters. Battle (1944) showed trauma in

"The experiences of the next second pass belief. You are wrenched and twisted and bounced about until you feel you will break. Around you you will see many of your fellows with eyes dangling out, with heads cut off, and with snapped spines. Quite a few of them will be smashed to death against the impeller blades.

1 22

Teleostean eggs for falls of only 0.6 meters, terminal velocity 240 cm/sec.

The next 4 to 5 seconds are spent in transit to the water box. Conduit conditions are equivalent to those ahead of the pump.

In the water boxes velocities increase to as much as 250 cm/sec. The irregular geometry of the boxes increases the turbulent intensity and accelerations of 10 g are common. Shear at the walls of the box can exceed 200 dynes/cm^2.

5 27

"The lot of you--the dead, the dying, and the living--will be spat out into comparative quiet. At least it will be no worse than it was before--although you will be.

"But, once again, you are carried along with a rush and the twisting and churning increases. You are spun and dropped. And the worst is yet to come.

The animal is forced into one of a multitude of 12-meter long, 2.3-cm I.D. condenser tubes. For the 2 to 8 seconds spent in the tube, heat absorbed from condensing turbine steam will raise the body temperature by some 11°C. The velocities remain quite high, 200 to 600 cm/sec. Turbulence induces

8 35

"Ahead is hell! Dozens of small openings lie before you and into one or another of them you go willy nilly. The twisting, rolling, and accelerations become even more intense. But the worst is the heat. It is worse than anything you can imagine or than you will ever experience in this world. Breathing! All you can do is

accelerations of 2 g to 14 g. Shear forces over 500 dynes/cm^2 exist near the tube walls. Biocide levels have dropped below 2 ppm and there is commonly an uptake of copper and other heavy metals from the tube walls.

gasp and wonder where your next breath is coming from--if you should live long enough to take it. And it goes on and on and on. About the only solace I have to offer is that the burning from the gas attack is easing off just a little. However, you won't find that much comfort since you will begin to taste copper, lead, and zinc in the water.

The second or so spent in the exit water boxes usually finds the organism subjected to the lowest absolute pressures of the journey, 0.3 atm, while other physical conditions are comparable to those found in the entrance water box.

"Again you will find yourself rushed from depth to surface and then flung into outer space. The sound of exploding bodies reaches a drumfire. Fragments and broken bodies are all around you. You will be fortunate if you are alive to see it, although it may not strike you that way. The shambles is worse than Pickerel's Charge at Gettysburg.

150 185

Exit conduits range in length from 100 to 500 meters and require transit times of 50 to 250 seconds. In this segment the mechanical conditions are similar to those in

"Then follows the long, long trek through the desert. The water is running more smoothly, but the heat and the suffocation go on for what seems like forever.

the upstream conduits. However, the temperature remains high, rarely falling more than 1°C while biocide levels fall to their exit values of 0.5 ppm.

The burning from the gas attack has definitely begun to abate.

How long the surviving creatures remain exposed to thermal and biocidal stresses after discharge ranges from seconds to hours depending upon whether a diffuser, multi-port jet, weir, or canal forms the outfall.

"At long last you're out--but not yet in the clear. All around you is the senseless carnage wrought by the monsters from outer space and you are feeling none too lively. Further, the brotherhood of fish is not a doctrine that applies at mealtimes. Many of your relatives and 'friends' will be gathered around to welcome you with gently smiling jaws.

An organism passing through a once-through cooling system of a power plant experiences a sequence of stresses: physical, thermal, and chemical. No answer can be given to the question, "How many organisms survive entrainment?" or to the question, "How can survival be maximized?" by evaluating a single stress. Even if a single stress were

"What practical advice can S 303 offer you? Like most theoretical courses, very little beyond, 'Don't get entrained.' However, when the bomb goes off it's always nice to know how the thing works.

limiting under a given set of conditions, there is no assurance that it will remain so when conditions are changed.

REFERENCES

American Nuclear Society. 1974. Entrainment: guide to steam electric power plant cooling system siting, design, and operation for controlling damage to aquatic organisms. ANS-18.3 Committee. Draft No. 8.

Battle, H. I. 1944. Effects of dropping on the subsequent hatching of teleostean ova. J. Fish. Res. Bd. Can. 6:252-256.

Coutant, C. C. and R. J. Kedl. 1975. Survival of larval striped bass exposed to fluid-induced and thermal stresses in a simulated condenser tube. Publ. No. 637, Oak Ridge National Laboratory, Oak Ridge, Tennessee (not a final report).

Morgan, R. P., R. E. Ulanowicz, V. J. Rasin, L. A. Noe, and G. B. Gray. 1976. Effects of shear on eggs and larvae of the striped bass, *Morone saxatilis*, and the white perch, *M. americana*. Trans. Am. Fish. Soc. 105:149-154.

GLOSSARY

ALEXIS STEEN

Acceleration. The time rate of change of velocity.

Acclimation. Becoming accustomed to a change in environment through behavioral, physiological or biochemical means. Also, Acclimatization.

Adaption. An alteration in the structure or function of an organism by which the organism becomes better fitted to survive in its environment.

Adenosine Triphosphate (ATP). A molecule that constitutes an immediate energy source for protein synthesis, cell division, muscle contraction, active transport across cell membranes, and bioluminescence.

Ambient Temperature. The temperature of the cooling water that would be observed if the plant were not operating. Also, Base Temperature.

Augmentation Pumping. See Dilution Water.

Base Temperature. See Ambient Temperature.

Bioassay. The determination of the potency of a substance by testing its effect on the growth or survival of an organism. See also, Thermal Bioassay.

Biocides. A class of substances lethal to organisms. Biocides are used in power plants to reduce slime accumulation on the condenser tubing.

Biomagnification. The increasing concentration of a pollutant especially metals, in organisms within one trophic level, or the increase with higher trophic level.

Blastula. An early stage in development of the fertilized egg in which the cells are arranged in layers and form a hollow sphere.

Blowdown. The portion of water in a closed cycle cooling system that is discharged to prevent an accumulation of dissolved solids.

Chloramines. The class of compounds formed by reaction between chlorine and ammonia or natural amino components.

Chlorine. The chemical element of atomic number 17, and atomic weight 35.453, often used as a biocide in the cooling water of power plants.

Chlorine Residual. The sum of the free available and combined chlorine.

Cleavage. The divisions of a fertilized egg until the number of cells is sufficient to form a blastula.

Closed Cycle Cooling. A cooling system which circulates all or a portion of the coolant. See Cooling Water.

Collision. Any interaction between free particles, aggregates of particles or rigid bodies in which they come near enough to exert a mutual influence, generally with exchange of energy. Does not necessarily imply actual contact. One cause of physical stress during pump entrainment.

Combined Available Chlorine. That portion of the chlorine present in water in combined forms as chloramines or other chloro-derivatives.

Community. A naturally occurring assemblage of organisms that live in the same environment.

Condenser. A device to reduce steam to liquid water by extracting the heat of vaporization.

Condenser Passage. See Pump Entrainment.

Cooling Tower. A structure for cooling water through partial evaporation. The steam to be cooled is broken into a fine mist to increase the surface area and promote cooling and then mixed with air brought in by natural convection currents or forced in by fans. The cooling water is re-cycled and only evaporative losses must be made up.

Cooling Water System. A cooling system where water is used as the coolant. It is composed of the intake structure, intake pumps, condenser, heat exchanger, and discharge structure.

Critical Thermal Maximum (CTM). The temperature at which locomotor activity becomes disorganized and an organism loses its ability to escape conditions which will promptly lead to its death. The CTM is determined by heating from acclimation temperature, at a constant rate that is fast enough to allow the deep body temperature to follow without significant time lag.

Debilitation. Decreased capability of an organism to function normally.

Delta T (ΔT). A difference in temperature. Measured in power plants from the ambient cooling water temperature to the heat exchanger (across the condenser).

Detritus. Particles of various sizes and shapes formed by the process of disintegration. Particles may have an inorganic or organic origin.

Dilution Water. Water that by-passes the condenser and is added to the cooling water before discharge to lower the temperature of the effluent.

Discharge Canal. An artificial channel to convey the cooling water from the discharge of a plant to the receiving water body.

Discharge Outfall. The end of the discharge structure.

Discharge Plume. A region where effluent can be delineated from the receiving waters by one or more of its characteristic properties. At a steam electric plant, the diagnostic property is usually temperature.

Dose Response. A response that is a function of the strength of the stimulus and the duration of its application. Thermal death in aquatic plants and animals has been shown to be a dose response; i.e. a function of temperature and the length of exposure.

Ecosystem. The biological community, its physical and chemical conditions and resources at a given location. Emphasis is on the obligatory relationships, interdependence and causal relations that form functional units. These units are composed of energy circuits, food webs, diversity patterns, nutrient cycles and evolutionary changes.

Effluent. That which is discharged. Usually refers to the discharge of cooling water at a power plant.

Entrapment. The capture of aquatic organisms in the intake canal or embayments due to water velocities, thermal gradients, or other causes.

Entrainable Size. That size which allows organisms to pass through the intake screens and to be entrained. The mesh size of screens is usually about 9-13 mm. See Entrainment.

Entrainment. The capture and inclusion of organisms into the cooling water of power plants. Two types of entrainment are recognized in power plants; plume entrainment and pump entrainment.

Entrainment Mortality. The death of organisms due to entrainment. See Entrainment Mortality Fraction and Entrainment Mortality Rate.

Entrainment Mortality Fraction. The ratio of the number of organisms killed to the number entrained in the system.

Entrainment Mortality Rate. The number of entrained organisms killed per unit time.

Estuary. A semi-enclosed coastal body of water which has a free connection with the open sea and within which sea water is measurably diluted with water derived from land drainage.

Excess Temperature. See Delta T.

Fouling Organisms. Organisms that attach themselves to ships, buoys, pilings, and other underwater structures. In power plants, fouling organisms may attach to the screens, condenser tube walls, and other parts of the cooling system and thereby impede flow, reduce heat transfer, and accelerate corrosion.

Free Chlorine. That portion of the chlorine concentration present as hypochlorous acid and/or hypochlorite ion.

Gas Bubble Disease (GBD). A pathological process due to gas supersaturation characterized by gas emboli, exophthalamus (pop-eye), systematic emphysema, disorientation and death.

Gas Emboli. The gas bubbles beneath the skin and within tissues and blood in fish due to gas supersaturation in surrounding waters. See Gas Bubble Disease.

Gastrula. The development stage of an egg in which invagination of the blastula occurs until the two walls meet.

Hatching Success. The percent of fertilized eggs that hatch.

Heat Shock. The physiological and behavioral reaction of aquatic organisms to sharp increases in temperature.

Heated Effluent. The effluent from electric power generating plants due to condenser or auxiliary cooling operations.

Ichthyoplankton. The fish egg and larval components of zooplankton.

Impingement. The striking or capture of aquatic organisms on the intake screens placed before the pumps to prevent the passage of organisms and debris through the plant.

Incipient Lethal Temperature. The temperature at which some specified fraction, usually 50%, of organisms acclimated to a particular temperature will die on continued exposure. Also, Lethal Threshold. See Upper and Lower Incipient Lethal Temperatures.

Inner-plant Mortality. The death of pump-entrained organisms due to mechanical, chemical and/or thermal stresses received inside a power plant.

Inner-plant Passage. Passage of organisms through the power plant, after pump entrainment and before discharge.

Jet Discharge. The high speed release of an effluent through an opening into the receiving medium. A jet discharge promotes rapid mixing and heat dissipation.

Lower Incipient Lethal Temperature. The temperature, below the acclimation temperature, at which 50% mortality occurs for an indefinite exposure time. See Incipient Lethal Temperature.

Mechanical Draft Cooling Tower. A device for cooling water in which the air flow, used as the coolant, is maintained by fans. Induced draft towers have fans mounted on top. Forced draft towers have fans located at the base. See Cooling Tower.

Meroplankton. Organisms, usually egg and larval forms of invertebrates and fish, which are planktonic during only a portion of their life cycles.

Mitosis. Somatic cell nuclear divisions occurring during growth and development.

Morphogenesis. The change in size and shape of bodily structures; e.g., tissues, organs, and systems.

Multi-Port Diffuser. An effluent discharge structure having more than one (usually many) openings. It promotes rapid mixing.

Natural Draft Cooling Tower. A device for cooling water through which air, used as the coolant, is circulated by the chimney effect of the tower shape. (Also, hyperbolic natural draft tower). See Cooling Tower.

Once-Through Cooling. A cooling system in which water is drawn from a source body, passed through the system and is returned directly to the contributing water body. Also Open Cycle.

Open Cycle. See Once-Through Cooling.

Ozone. A form of oxygen, O_3, ozone is a blue gas formed when ordinary oxygen is subject to electrical discharge.

Phytoplankton. Planktonic organisms which are members of the plant kingdom. See Plankton.

Plankton. Those aquatic organisms capable only of weak swimming or passive floating in the water; primarily microscopic.

Physical Abrasion. The physical rubbing, scraping, or colliding of solid objects. At power plants, pump-entrained organisms may suffer abrasion.

Plume Entrainment. The inclusion of aquatic organisms into the discharge plume along with the receiving water without passing through the plant. See Discharge Plume.

Population. A group of organisms of the same species in a given environment.

Post-Yolk Sac Larvae. The stage of fish larval development after the yolk sac has been absorbed by the larval body.

Pressure. The force per unit area exerted perpendicular to the surface.

Pump Entrainment. The entrainment of aquatic organism by the intake pumps.

Receiving Water. The body of water to which effluents from the power plant are discharged.

Salinity. The concentration of total dissolved solids in a sample of water; usually expressed in parts per thousand o/oo (g/kg).

Seston. Suspended particulate matter, living or dead.

Shear Stress. Force per unit area acting in a direction tangent to a surface.

Source Water. Water body from which water is drawn.

Square Wave. A wave form characterized by vertical discontinui-nuities. This applies to time-excess temperature histories where the organism is subjected to an instantaneous temperature rise and subsequent decrease. See Time-Excess Temperature History.

Straight-Hinge Larvae. The last stage of planktonic shellfish larvae prior to settlement on the sea bottom.

Sublethal Effects. Physiological stresses which do not cause immediate death, but which reduce an organism's ability to reproduce, grow, and survive.

Tempering Water. See Dilution Water.

TL_{50}. The temperature at which 50% of the test organisms are able to survive for a specified period of exposure, usually, 24, 48, or 96 hours.

Temperature Tolerance. The ability to endure a given temperature, often calculated by the Critical Thermal Maximum method. See Critical Thermal Maximum.

Thermal Bioassay. A determination of the effect of changes in temperature upon the viability of an organism. See also Bioassay.

Thermal Resistance Curve. A graph delineating the conditions for survival (usually at the 50% level) for organisms exposed to different temperature-time combinations.

Time-Excess Temperature History. The changes in temperature with time to which entrained organisms are exposed at an operating power plant.

Traveling Screen. Screen placed before the intake pumps that moves either vertically or horizontally. Material caught on the screen is usually passed through a sluice-way for removal.

Trochophore. The free-swimming, ciliated, larval stage of shellfish.

Turbulence. An irregular fluid flow whose velocity field is a random function of space and time. Such flows are characterized by advective transports which are orders of magnitude larger than the corresponding molecular transports.

Ultimate Incipient Lethal Temperature. The highest incipient lethal temperature that can be achieved by acclimation. See Incipient Lethal Temperature.

Weir. A barrier constructed to divert the flow of water.

Xenobiotic. Those compounds which do not occur in nature. Also, exotics.

Zooplankton. Those planktonic organisms that are members of the animal kingdom.

[illegible]

[illegible] lethal temperature.

[illegible] used to divert the flow of water.

Xenobiotics: Those compounds which do not occur in nature.

Zooplankton: Those planktonic organisms that are members of the animal kingdom.

INDEX

D

E

F

G

H

I

S

T

A
B
C 8
D 9
E 0
F 1
G 2
H 3
I 4
J 5